Ceramics in Energy Applications

New Opportunities

Ceramics in Energy Applications
New Opportunities

Proceedings of the Institute of Energy Conference held in Sheffield, UK, 9–11 April 1990

Adam Hilger, Bristol and New York

British Library Cataloguing in Publication Data

Ceramics in energy applications
  1. Energy conversion systems. Use of ceramics
  I. Institute of Energy
  621.042

ISBN 0-7503-0035-3

Library of Congress Cataloging-in-Publication Data are available

Published under the Adam Hilger imprint by IOP Publishing Ltd
Techno House, Redcliffe Way, Bristol BS1 6NX, England
335 East 45th Street, New York, NY 10017-3483, USA

Printed in Great Britain by J W Arrowsmith Ltd, Bristol

Contents

vi   *Contents*

†Paper not available at time of going to press.

Session 5: New Applications and Developments

CERAMICS IN ENERGY APPLICATIONS
New Opportunities

THE INSTITUTE OF ENERGY

and associated societies overseas:

American Society of Mechanical Engineers
Australian Institute of Energy
Canadian Institute of Energy
L'Institut Francais de l'Energie
Fuel Society of Japan
Verein Deutscher Ingenieure (GET)

and in association with

The Institute of Ceramics
British Ceramic Research Ltd
The Institution of Refractory Engineers
The Institute of Metals
The Institution of Chemical Engineers

CERAMICS IN ENERGY APPLICATIONS
New Opportunities

Conference Organization

Professor B J Brinkworth

President
The Institute of Energy

Organizing Committee

Mr M L Hoggarth
(Chairman—organizing committee)
Mr N G Worley
Dr E Hampartsoumian
Mr M Mason
Mr D F Hibberd
Dr N Fricker
Dr G Padgett
Professor B Wilshire
Mr D Bate

British Gas plc
The Institute of Energy
University of Leeds
The Institution of Refractory Engineers
British Steel Corporation
British Gas plc
British Ceramic Research Ltd
University College Swansea
ICI Engineering

The Institute of Energy

Mr C Rigg
Ms J A Higgins

The Secretary
Conference Manager

Paper presented at Conf. on Ceramics in Energy Applications, Sheffield, April 1990
Session 1

The performance of materials used for ceramic radiant and immersion tubes

A.J. Jickells, S. Matthews, P. Sihre

Midlands Research Station, Wharf Lane, Solihull, West Midlands.
B91 2JW

ABSTRACT: The past 5-10 years has seen a number of developments in indirect gas-fired heating systems. Two such systems, namely radiant (700-1250°C) and immersion tubes (450-900°C) have recently exploited the higher temperature capabilities of ceramics. The tubes are fabricated from silicon nitride bonded silicon carbide (SNBSC), however, during field trials of both applications initial problems were encountered with variable tube lives. Laboratory studies and examination of used tubes confirmed that the principle cause of failure was oxidation of SNBSC. Service performance is a function of the tube manufacturing route and further work is concentrating on adjusting the base material to achieve a more consistent product with improved oxidation resistance.

1. INTRODUCTION

In many applications of industrial gas heating, there is a requirement to control the furnace atmosphere in order to preserve or modify the properties of the material being heated. At the same time there is a need to maximise the thermal efficiency of the heating process to minimise operating costs. To meet these demands gas fired radiant tubes with waste heat recovery have been developed. In many low temperature applications, metallic tubes are used, but at higher temperatures and in more hostile environments ceramic tubes are required.

The Midlands Research Station of British Gas has developed ceramic radiant tubes for the indirect heating of furnaces with controlled atmospheres and ceramic immersion tubes to heat molten zinc and aluminium with the tube immersed in the molten metal. After initial development, licensees have produced and marketed commercial versions. The introduction of the technology into a number of different applications (Table 1) was conducted via production field trials during which the performances of all aspects of the technology could be monitored and any problems identified and rectified. During two of these field trials, initial problems were encountered with variable tube life. This paper describes the identification of the cause of these failures and the improvements which have subsequently led to economically and technically acceptable tube lives being obtained in these applications.

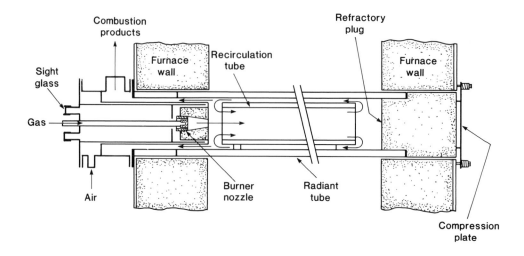

FIG. 1. SCHEMATIC OF CERAMIC RADIANT TUBE

2. TUBE DESIGN AND OPERATION

The gas fired ceramic radiant tubes have been developed primarily for operation in the temperature range 900°C to 1250°C. The tube design, Figure 1, consists of an open ended outer (radiant) tube and an inner (recirculation) tube fitted with a small recuperative burner at one end and a ceramic plug at the other. Each end of the tube is supported in the furnace wall using robust mounting techniques which accommodate thermal expansion and reduce tensile stresses in the material. The recuperator pre-heats the incoming air prior to mixing with the fuel gas at the burner nozzle. Combustion takes place inside the tunnel which is designed to provide a high velocity jet of combustion products to promote recirculation around the inner tube. This gives a uniform temperature along the outer tube.

The outer tube, which has a diameter of 170 mm and working length of 1.4 metres, can dissipate up to 40 kW/m² of heat from the radiating surface with the outer tube surface temperature upto 1350°C and the inner tube at a higher temperature of up to 1450°C. The inner tube and the inside surface of the outer (radiant) tube are therefore exposed to the combustion flue gases ($72\%N_2$, $17\%H_2O$, $9\%CO_2$, $2\%O_2$) whilst the outer surface of the radiant tube is exposed to the furnace atmosphere. The thermal efficiency based on the gross calorific value of the fuel is 50 to 55% with the use of a recuperative burner. In practice this resulted in improvements of between 25% and 50% of the thermal efficiency of the plant compared to the previous equipment.

The ceramic immersion tube is of a similar design, except that a closed end tube is used. The burner is mounted vertically facing downwards into the tube. The tube is immersed in the molten metal and the outside surface of the radiant tube is exposed to three environments, air in the upper portion, a zone incorporating a flux/melt/air interface and a zone of molten metal.

3. TUBE MATERIAL SELECTION

Since tubes were intended for use in range of corrosive and oxidising environments, there were limitations on types of material which could be considered. The combination of these external conditions and the high operating temperatures required the careful selection of a non-oxide ceramic with a high thermal conductivity in order to maximise thermal efficiency and thermal shock resistance. The strength of the material was not a major consideration. Finally the material had to be capable of being formed into large, tubular components at an acceptable cost.

The material selected as most closely meeting these specification requirements was silicon nitride bonded silicon carbide, SNBSC.

4. TUBE MANUFACTURE

The tubes are manufactured from a controlled mixture of coarse, medium and fine silicon carbide grains together with a controlled size of silicon metal powder and green and permanent binders. The mixture is then shaped into the green tube. The method of shaping varies with manufacturer; two manufacturers have provided the vast majority of the tubes using specific processing routes of either manual pressing (Manufacturer 1) or isostatic pressing (Manufacturer 2). After drying, the "green" tube is hand ground to produce a relatively smooth surface finish. The final stage of manufacture involves firing the tube in a nitrogen atmosphere at around 1400°C. The silicon metal reacts with the nitrogen to produce a continuous bonding phase of silicon nitride.

The finished material from both manufacturers contains about 75% SiC, 23% Si_3N_4 with a residual porosity of around 15%.

5. LABORATORY STUDIES

Two investigational routes were followed. A programme of laboratory testing of sample materials was initiated and studies were also made on materials from used radiant and immersion tubes.

5.1 Comparison of Unused Material

An examination of the unused material shows differences in the microstructure of the materials produced by Manufacturers 1 and 2. In both cases the major phase is silicon carbide. The distribution of the fine, medium and coarse silicon carbide grains is reasonably even although agglomerations of similarly sized particles have been observed in both materials. The bonding phase of Material No. 1 is a mixture of silicon and cristobalite, SiO_2, whereas the bond of Material No. 2 consists of silicon nitride and silicon oxy-nitride, Si_2ON_2. The cristobalite present in Material No. 1 has been measured at levels of 5% to 8% in an unused sample (Fig 2). Optical microscopy shows that it is present principally around pore sites. It is less easy to distinguish between silicon nitride and silicon oxy-nitride in Material No. 2. There are small pockets of unreacted silicon metal in the bonding phase of both materials. The mean density of each material was determined (from fifty, one centimetre cubes). A value of $2.63 gcm^{-3}$ was obtained for both materials, but the standard deviation of the cubes from Material No. 1 was three times that of Material No. 2. The crush strengths of the unused materials are very similar at around 220-240MPa.

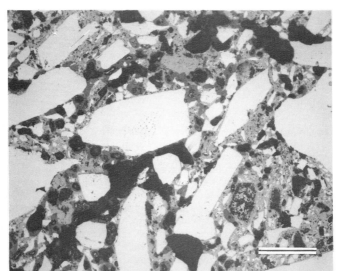

Bar width
= 100 μm

FIG. 2. MATERIAL No. 1; AS RECEIVED

5.2 Oxidation Studies

Test cubes ($1cm^3$) were cut from unused tubes produced by Manufacturers 1 and 2. The materials were exposed to air and combustion product environments for 500 hours at temperatures in the range 800°C to 1200°C (ten cubes for each test condition). The material property changes were examined in terms of weight and volume changes, x-ray diffraction (XRD), fourier transform infrared spectroscopy (FTIR), optical microscopy, scanning electron microscopy (SEM) and cold crush strength. In addition some limited Thermo-gravimetric analysis (TGA) scans were conducted on the material.

5.3 WEIGHT CHANGE

Isothermal oxidation tests (Table 2) show that in both air and
combustion products Material No. 2 experiences significantly higher
weight gains than Material No. 1. This holds for every test
condition. Material No. 2 also shows markedly higher weight gains in
combustion products (900-1150°C) compared to air. The same effect is
observed in Material No. 1 above 1000°C. Weight changes were
estimated from ten samples thus a spread of individual weight gains
was obtained. In general the weight gains for Material No. 1 were
subject to quite a large scatter particularly, for example, at 950°C
in air. The effect of temperature on weight change (Fig. 3) reveals
quite different behaviour in Materials No. 1 and 2. At the lowest
temperature, 800°C the weight changes are only significant in Material
No. 2. As the temperature is increased, however, the weight gains
gradually rise, and in the case of Material No. 2 peak at 900°C in air
and 1000°C in combustion products. Beyond these temperatures the mean
weight gains fall off. The oxidation behaviour of Material No. 1
is not characterised by such readily discernible peaks. Instead the
weight gains are seen to increase steadily throughout the range of
temperatures investigated.

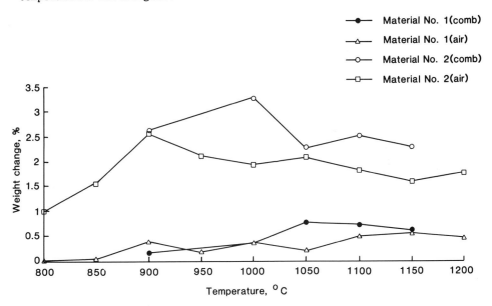

FIG. 3. WEIGHT CHANGE VS. TEMPERATURE
LAB TESTS IN AIR AND COMBUSTION PRODUCTS

5.4 Volume Change

Contrary to the weight change data the volume changes (Table 2) are
much the same for both materials. Oxidation to silica is accompanied
by a volume expansion which occurs because the silica has a lower

density, 2.20gcm^{-3} than the silicon carbide/silicon nitride, 3.20gcm^{-3}. However, measurements would be influenced by surface irregularities, e.g. scales and protrusions. In particular above 1100°C material No. 2 was covered in a surface glaze which affected the measurements. Material No. 1 was also covered in a surface deposit, but this was quite different in nature to a glazing effect.

5.5 FTIR Analysis

FTIR analysis (Table 2) was carried out on one cube from each test regime. This technique has been specially adapted to provide a quantitative assessment of cristobalite, SiO_2 in samples of SNBSC. As cristobalite is the principal oxidation product the degree of oxidation can be defined by this technique. Tests on the unused material indicate that whilst Material No. 2 does not contain any cristobalite, Material No. 1 contains quite significant quantities (5-8%). The levels of cristobalite production in Materials No. 1 and 2, as a result of oxidation, are reasonably similar, although there is greater variability in Material No. 1. The cristobalite content of Material No. 2 increases with the test temperature between 800°C and 1050°C. Above 1050°C the cristobalite levels fall off slightly.

5.6 XRD Analysis

XRD results (Table 2) confirm that cristobalite is produced in both materials over the range of test temperatures. It is not clear which species (Si, SiC, Si_3N_4, Si_2ON_2) have been oxidised as there are only very minor phase differences between samples. The technique is semi-quantitative and can therefore only differentiate between unoxidised and oxidised samples.

5.7 Crush Strength

In these tests oxidation appears to strengthen the material (Fig. 4). All of the test samples are stronger than the unused material. There is an increase in strength as the test temperature increases from 800°C in air. The strength peaks at 900°C (Material No. 1) and 1050°C (Material No. 2). These effects are less prominent after tests in combustion products.

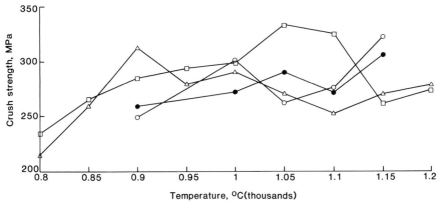

FIG 4   CRUSH STRENGTH VS TEMPERATURE LAB TESTS IN AIR AND COMBUSTION PRODUCTS

5.8 Optical Microscopy

Oxidation could be observed in nearly all of the test samples examined using optical microscopy. Material No. 1, even in the as received condition, has a high proportion of the dark silica phase around pores (Fig.2). Similarly there are high proportions of silica in all of (Material No. 1) samples exposed to combustion products and high temperature (>1050°C) air. Where large amounts of silica have formed it is found to extend well into the bonding phase and not just be associated with pores (Fig. 5). In Material No. 2, very little silica could be detected in samples tested below 900°C in air. Between 900°C and 1100°C small amounts of silica are formed in the bonding phase near the surface. Above 1100°C the oxidation manifests itself in the form of a surface glaze of silica (Fig. 6). This behaviour is duplicated in air and combustion products. Closer examination of the oxidation product, in both cases, revealed that it consisted of more than one phase indicating that other species play a part in the oxidation process.

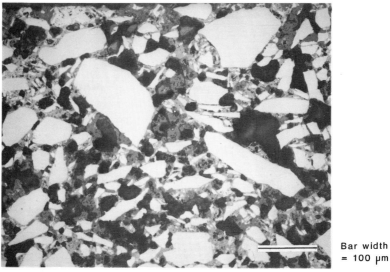

Bar width
= 100 μm

FIG 5 No.1: 500 HOURS AT 1150°C (AIR)

Bar width
= 100 μm

FIG. 6. MATERIAL No. 2; 500 HOURS AT 1150°C (AIR)

5.9 Scanning Electron Microscopy

SEM examination of the test samples using backscattered electron imaging reveals that the oxidation product is mainly pure silica with some phases containing Si, Ca, Fe and K oxides (possibly in the form of silicates). Ca and Fe are more prevalent in Material No. 1 whereas Ca and K are found in Material No. 2.

5.10 Thermogravimetric Analysis

Studies using TGA (Fig. 7) confirmed the temperatures at which oxidation occurred in these materials. This starts at temperatures above 700°C and the maximum rates are seen at 1100°C (Material No. 2) and 1200°C (Material No. 1).

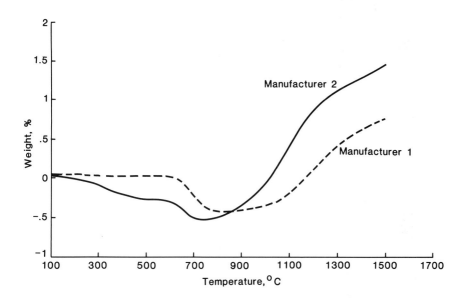

FIG.7. THERMOGRAVIMETRIC ANALYSIS OF SILICON NITRIDE BONDED SILICON CARBIDE

6. TUBE PERFORMANCE

6.1 Radiant Tube Field Trial Data

In an application involving a continuous kiln cracking was noted on a number of tubes (from Manufacturer No. 1) after a relatively short operating life (2 months). Cracking was most prevalent in the 960°C temperature zone of the kiln (Figs. 8 & 9). Tubes in other zones of the kiln were not as susceptible to this form of cracking. When these tubes were replaced, they were returned to the Midlands Research Station for examination. Samples were examined using optical and scanning electron XRD and FTIR techniques and the strength of tube materials were determined using crush strength on cubes cut from the tubes at positions free of surface defects.

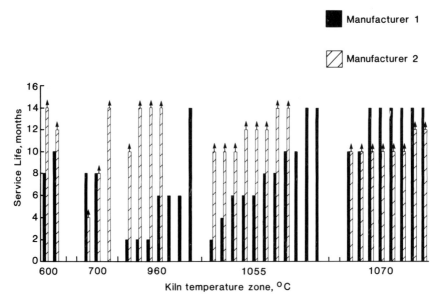

FIG. 8. SERVICE LIVES OF RADIANT TUBES
(CONTINUOUS KILN APPLICATION)

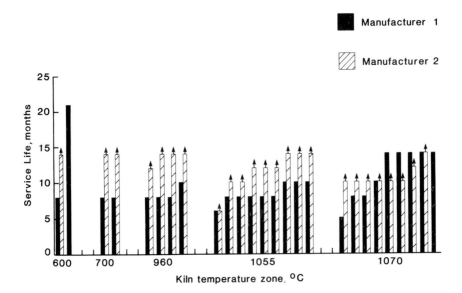

FIG. 9. SERVICE LIVES OF INNER TUBES
(CONTINUOUS KILN APPLICATION)

Tubes from both Manufacturers No. 1 and 2 were used in the kiln and those from Manufacturer No. 2 have achieved consistently longer lives, particularly in the 960°C zone. Many of the tubes from Manufacturer No. 2 are still used in the kiln, a number of tubes having exceeded 14 months continuous operation at 960°C, 1055°C and 1070°C (Figs. 8, 9).

Amongst outer tubes from Manufacturer No. 1, oxidation to silica (in the form of cristobalite) could not be detected below 700°C but was most severe at 960°C (Table 3). At 1070°C the levels of oxidation were much smaller. Inner tubes, which operate at higher temperatures than the outer tubes, were found to have cristobalite contents ranging from 2-47%. Oxidation of inner tubes appeared to be much less sensitive to temperature and service life compared to the outer tubes.

The crush strengths of used tube material from Manufacturer No. 1 varied greatly. A few tubes retained their original strength (e.g. at 1055°C and 1070°C). Most tubes were significantly weaker, particularly at 960°C and 700°C. Tubes from the 1055°C and 1070°C zones were on the whole only slightly weakened. Material from the 600°C zone had also been weakened. In general inner tubes retained their strength better than the radiant tubes, even with high cristobalite contents.

The service life of Material No. 2 has been shown to be significantly better than Material No. 1 (Figs. 8 & 9). The limited evaluation of Material No. 2 (Table 3) indicates that the material is susceptible to quite high levels of oxidation. An inner tube from the 1055°C zone has undergone a substantial reduction in strength. Material from the 1070°C zone, however, retains its strength after 14 months service.

Examination of the fracture surfaces of a number of used tubes from Manufacturer No. 1 showed that the fracture paths changed as the levels of oxidation increased (Figs. 10a and b). In lightly oxidised (or unused) samples the fracture was "transgranular"; through both the bonding phase and silicon carbide grains. In heavily oxidised samples the fracture path changed to "intergranular"; fracture was primarily through the bonding phase, very little fracture of silicon carbide grains occurred. The transition from "transgranular" to "intergranular" was accompanied by a significant reduction in strength. Laser induced ion mass analysis (LIMA) indicated that the "intergranular" fracture path was mainly through silica.

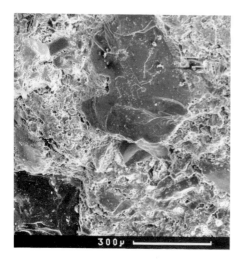

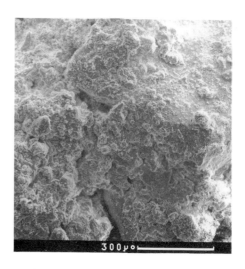

FIG. 10a. TRANSGRANULAR FAILURE IN
UNUSED TUBE MATERIAL

FIG. 10b. INTERGRANULAR FAILURE IN
USED TUBE MATERIAL

7. IMMERSION TUBE FIELD TRIAL DATA

Immersion tubes operate in a very different environment from radiant tubes. Whilst the flue gas environment on the inside of the tube is the same, the outside of the tube is exposed to three environments; the top of the tube is exposed to air, below this the tube is exposed to melt/flux line and the lower 70% of the tube is immersed in the molten metal.

Service experiences indicate that tube lives of greater than 12 months are routinely obtained in zinc baths. In the field trial of an aluminium melter problems were encountered with unpredictable and in many instances short tube lives, Fig. 11. An examination of failed tubes indicated that failure occurred in all instances below the melt line usually near the bottom of the tube. As was the case with the radiant tubes, differences were seen between the lives achieved by tubes produced by different manufacturers.

Examination of the failed tubes again indicated that oxidation was occurring within the tube wall. This resulted in an increase in silica content, a decrease in the material strength and a change in fracture morphology, ref. 1. An additional, important factor was found. The molten aluminium neither wetted silicon carbide nor the freshly manufactured silicon nitride bonded silicon carbide. As the oxidation proceeds, however, silica is formed adjacent to the molten aluminium. The aluminium reacts with the silica forming alumina (corundum) and silicon.

$$4Al + 3SiO_2 = 2Al_2O_3 + 3Si \qquad \ldots\ldots\ldots(1)$$

and may also react with the Si_3N_4

$$4Al + Si_3N_4 = 4AlN_4 + 3Si \qquad \ldots\ldots\ldots(2)$$

Microstructural and analytical evidence was found for reaction (1), no such evidence was found for reaction (2). This may be because the reaction is kinetically slow or because oxidation occurs preferentially.

Reaction (1) leads to a rapid degradation of the silicon nitride bond and a marked decrease in strength (refs. 1, 2).

It is further clear that the presence of SiO_2 in the as supplied material will be highly disadvantageous since this will enable a reaction to initiate at the start of service rather than once oxidation has occurred. Therefore the presence of SiO_2 in variable amounts in the as supplied tubes from Manufacturer No. 1 was a contributing factor to the short and unpredictable immersion tube lives.

The change of material to Manufacturer No. 2 and the use of zircon coatings on the inside of the tubes were both shown to lead to marked improvements in tube life (Fig. 11).

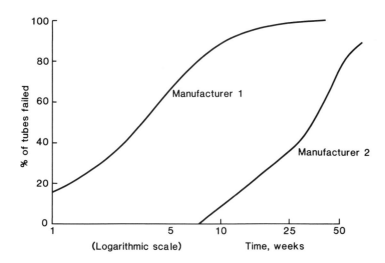

FIG. 11. CERAMIC IMMERSION TUBE SERVICE LIVES

8. DISCUSSION

The manufactured tube materials show some significant difference. Material from Manufacturer 1 has a significant level of cristobalite and this level varies between different samples analysed. The cristobalite is present around pores in the material. Material from Manufacturer 2 has no cristobalite present in the as received condition. The density of this material is much more consistent, reflecting the more uniform fabrication process. Finally there are differences in the nature and level of trace elements present in the material.

The oxidation weight change behaviour of Material No. 2 (Fig. 3) may be attributed to the previously observed behaviour of silicon carbide and silicon nitride over a range of temperatures (refs. 3 & 4). At high temperatures (>1100°C) it has been reported that silicon carbide and silicon nitride can form protective "glazes" of silica which effectively seal off the rest of the material from further oxidation. In this situation the rate controlling step for further oxidation is the diffusion coefficient of the oxidant through this layer of silica. At lower test temperatures (900°C–1050°C) a protective layer of silica does not readily form and surface pores remain open thus providing a short circuit path for the diffusion of the oxidant. A larger surface area is hence available for oxidation resulting in the higher weight gains at these temperatures. The weight change behaviour of Material No. 1 cannot be defined in terms of the above arguments. The apparent increase in oxidation at higher temperatures is associated with the fact that a protective glaze does not form in this material.

The increase in room temperature strength (Table 1, Fig. 4) of the oxidised test samples may be due to the influence of oxidation on pore sizes within the material (ref. 2). At lower test temperatures oxidation can occur at internal pore sites and cracks due to the short circuiting effect discussed previously. Growth of silica leads to a decrease in apparent defect size. If the pores are treated as flaws it is possible to apply standard fracture theory. The silica formed at pore sites may act in two ways.

i) Due to the volume increase on oxidation, silica partially fills the cracks and pores, thus reducing their size and blunting them.

ii) Increase the surface energy of the pore.

Both effects increase the yield strength, given by:

$$\sigma_f = \frac{1}{A} \cdot (2E \quad)^{1/2}$$
$$\text{(C)}$$

where σ_f = yield strength, Nm^{-2}
A = geometrical factor (constant)
E = Young's modulus, Nm^{-2} (constant)
$\quad$ = surface energy, Jm^{-2}
C = flaw size, m

The strengthening effect is diminished at higher test temperatures as the oxidation occurs principally at the surfaces. Oxidation of internal pores is thus minimised and there is no apparent decrease in flaw size.

This strengthening mechanism reaches a plateau once the cracks and pores are sealed. This plateau is reached more rapidly in the material from Manufacturer 1 and at a much lower silica level. It is thought that this is due to differences in the two materials. The pores in material from Manufacturer 1 are already filled with silica and only oxidation of cracks occurs. In material from Manufacturer 2 oxidation of both pores and cracks occurs.

In contrast to the 500 hour laboratory tests, radiant and immersion tube service lives are in the region of a few months to well over a year.

The long term effects of oxidation are observed in both the radiant and immersion tubes and there are parallels which may be drawn with the laboratory tests. Examination of used radiant and immersion tubes indicate that the most severe oxidation occurs at temperatures between 900°C and 1055°C (Figs.8 & 9). These temperatures correspond to those at which internal oxidation (at pore sites) is observed in the laboratory test samples. It is thought that over longer periods, the extent of internal oxidation becomes significant and the strength decreases due to oxidation of the silicon nitride bonding phase to the weaker silica phase. Silica levels of 25-40% can only be accommodated as an almost continuous phase throughout the material, and not merely at pore sites and external surfaces. At 1070°C, however, the radiant tubes are found to have much longer service lives and the levels of oxidation are comparatively small, even after 14 months service. This temperature approaches that at which "protective" oxidation can take place (>1100°C).

The consistency of Material No. 2 is observed in both laboratory tests and field trials and its behaviour agrees with previous silicon carbide and silicon nitride oxidation theories (refs. 2-4). Material No. 1, on the other hand is prone to unpredictable behaviour in laboratory tests and service. There are probably three key factors which determine the oxidation behaviour of Material No. 1;

   i)   Variations in the density of the unused material (due to the pressing technique).

   ii) The presence of silica in the unused material.

   iii) Impurities, originating from binders, which take part in the oxidation process.

Variations in the density of the material imply a corresponding variation in porosity and this will lead to uneven oxidation in service. More porous zones will be more likely to oxidise and the accompanying volume expansion will be isolated to these zones. The stresses generated by the volume expansion and the concurrent loss of strength due to bond oxidation account for cracking observed in radiant tubes of Material No. 1 working in the temperature zones 960°C and 1055°C.

Thus the difference in weight change on oxidation (Fig. 3) can be attributed to the absence of silica in pores in material from Manufacturer 2. It is proposed that oxidation of pore volume is not detrimental to strength. Once the pores are sealed, oxidation proceeds through the bonding phase, possibly at silicon nitride/silicon carbide interfaces leading to a decrease in strength. The silica in the unused material may also act as seeding sites for oxidation at high temperatures (>1100°C) thus limiting the ability to form a protective glaze.

The surface layer of silica formed by Materials 1 and 2 contains different impurities (from SEM studies). It is difficult to assess their exact role from our studies, however, the presence of Fe and Ca (Material No. 1) and Ca and K (Material No. 2) may affect the viscosity and diffusional transport properties of the oxide layer at pore sites and on the surface of the material.

Material No. 2 has shown itself to be the most suitable for the continuous kiln, radiant tube application and the aluminium melting, immersion tube application. All of the evaluation indicates that tube performance is largely dependent on the consistent quality and specification of tubes. If these requirements are met the service performance can be modelled using generally accepted oxidation theories.

9. FUTURE DEVELOPMENTS

9.1 Coating Studies

Historically some tubes, particularly immersion tubes, have been given a proprietary wash, one of which consists of silicon metal and borax ($Na_2O.2B_2O_3$). At typical operating temperatures i.e. above 900°C this coating has a low viscosity and wets the tube surfaces thus forming a barrier against oxidation. In this condition the glaze is liable to run and this may cause problems in immersion tubes. This type of coating did not provide adequate protection against oxidation and tube lives were not increased.

British Gas has investigated an alternative coating material, zircon, which has many of the desired properties for this application, i.e.

  i)   Similar thermal expansion coefficient.
 ii)   Good thermal conductivity.
iii)   High emissivity which can be increased using additives.

The coating is applied, in the form of a water slurry, by painting or dipping. Unfortunately it does not fuse easily (due to its refractoriness) and a continuous layer is difficult to form and liable to spall or crack. Laboratory tests (Fig. 12) have shown that the coating is effective in reducing oxidation (ref. 1).

Experience with zircon coats in service has shown that immersion tube performance can be significantly improved. The performance of zircon coated radiant tubes is generally improved only slightly. Complete elimination of oxidation is not possible as it may still occur by diffusion of oxygen through the coating as well as through areas of poor adhesion and cracks in the coating. Further work in this area is continuing.

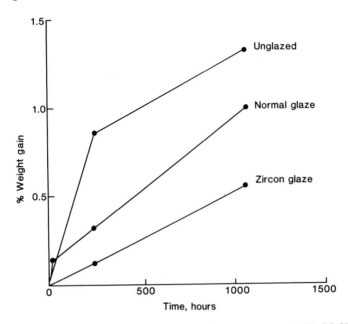

FIG. 12. EFFECT OF COATINGS (GLAZES) ON THE OXIDATION OF SNBSC

9.2 Tube Development

A collaborative programme was initiated with Manufacturer No. 2, to determine the effect of fabrication variables upon oxidation resistance; with the ultimate aim of obtaining a better product and being able to specify materials properties for British Gas applications of radiant and immersion tube technology.

From the materials aspects the aims were to increase bulk density and reduce the porosity and gas permeability of SNBSC. It is thought that these goals could be achieved by incorporating various adjustments into the standard mix. These adjustments comprised of changes to the silicon carbide and silicon metal particle size distributions and alterations to the mixing and forming processes.

The manufacturer produced a number of test samples for this programme which were supplied to British Gas for evaluation. These included tube specimens isopressed to 300mm length, pressed bars and QA test pieces. Samples were evaluated by the manufacturer for apparent porosity, bulk density and modulus of rupture. British Gas used their test samples for a detailed examination of the microstructure, chemistry and oxidation behaviour of each material.

Three 13 x 1 x 1 cm^3 bars were cut from each tube which were used for bulk density measurements and oxidation tests. Oxidation tests were conducted at 1050°C in air and combustion products for 500 hours.

A similar routine of evaluation was used including FTIR, XRD, SEM, optical microscopy and crush strength. The programme was designed to allow maximum feedback of results to the manufacturer, so that adjustments could be made to the standard mix on an iterative basis.

In order to obtain some long term oxidation data, tube samples (150mm long), were placed in a radiant tube kiln at Midlands Research Station. The kiln, which is currently used for performance and life testing of radiant tubes, has been running at temperatures between 900°C and 1050°C. Eight of the development samples have already received six months exposure within this temperature range.

The development programme with Manufacture No. 2, whilst as yet incomplete, has produced some significant improvements to the standard material in terms of both bulk density and oxidation resistance.

10. SUMMARY

In summary therefore, experience in the field has demonstrated gas-fired ceramic radiant and immersion tube heating technology as an efficient means of heating furnaces and liquid metals. In all cases it has been possible to demonstrate that tubes fabricated in silicon nitride bonded silicon carbide can give reliable long term service lives.

It has been found that in certain applications, tube degradation can occur as a result of oxidation of the silicon nitride bonding phase. The susceptibility of the material to such degradation is dependent on the method of tube manufacture and with care the problem can be overcome. It is important that the nitriding process is closely controlled to minimise the presence of silicon/silica in the as manufactured tube, to control the presence of minor impurities which affect the ability of the material to form a protective surface silica layer during service and to ensure uniform and repeatable structures throughout the tube. It has been demonstrated that it is possible to further inhibit oxidation by the application of ceramic coatings.

11. REFERENCES

1. BURKE, P. and .WEDGE, P.J., "Ceramic Immersion Tubes for Melting and Holding", Foundry Trade Int., Jun/Feb. 1988.

2. JOHNSTON, M.W., LITTLE, J.A. "Degradation of Oxidised SiC-Si$_3$N$_4$ in Molten Aluminium", Mats. Sci. Tech. In press.

3. MAEDA, M., NAKAMURA, K. and TSUGE, A. "Scale Effect of the Testing Furnaces on the Oxidation of Silicon Carbide Ceramics". J. Mats. Sci. Lett 8(1989) 144–146.

4. MAEDA, M., NAKAMURA, K. and TSUGE, A. "Scale Effect of the Testing Furnaces on the oxidation of Silicon Nitride Ceramics". ibid, 8(1989) 195–197.

TYPE	APPLICATION	OPERATING SERVICE TEMPERATURE	No. OF TUBES IN EACH UNIT	TYPICAL TUBE LIVES / months	No. OF UNITS IN SERVICE
RADIANT	Bar end and billet heating	1230 C	2	12	1
RADIANT	sintering	1090 C	1 – 2	14	2
RADIANT	Heat treatment	900 – 1100 C	1 – 3	12*	4
RADIANT	Ceramic tiles (continuous tunnel kiln)	850 – 1070 C	24	12 min.	1
IMMERSION	Zinc melting	460 – 500 C	2	6 – 12	5
IMMERSION	Galvanising	460 C	6	6	18
IMMERSION	Aluminium holding	730 – 750 C	1	6	11
IMMERSION	Aluminium melting	750 C	8	6	1

TABLE 1 DETAILS OF APPLICATIONS OF RADIANT AND IMMERSION TUBES

Material No.1

TEMPERATURE /degrees C	ENVIRONMENT	WEIGHT CHANGE /% (s.d.)	VOLUME CHANGE /% (s.d.)	XRD RESULTS cristobalite		FTIR RESULTS /% cristobalite	CRUSH STRENGTH /Mpa (s.d.)
				low	high		
UNUSED	--	--	--	m		5 - 8	239 (21)
800	AIR	0.02 (0.02)	0.08 (0.09)	m	t / s.o.	11.3	215 (79)
850	--	0.04 (0.13)	0.21 (0.19)	s.o. / m	t. / s.o.	9.0	259 (49)
900	--	0.39 (0.52)	0.35 (0.18)	m	t / s.o.	4.8	313 (52)
950	--	0.19 (1.46)	0.54 (0.17)	s.o.	t / s.o.	10	279 (18)
1000	--	0.37 (0.36)	0.33 (0.32)	m	t / s.o.	5.9	290 (27)
1050	--	0.27 (0.19)	0.32 (0.16)	m	t / s.o.	17	270 (71)
1100	--	0.50 (0.22)	0.53 (0.11)	m	t / s.o.	6.7	252 (35)
1150	--	0.56 (0.24)	1.64 (0.50)	m	t / s.o.	10.5	270 (30)
1200	--	0.46 (0.13)	void	s.o. / m	t / s.o.	11.5	278 (26)
900	combustion-products	0.18 (0.30)	0.42 (0.35)	s.o. / m	t / s.o.	9.0	259 (47)
950	--	0.03 (0.11)	0.74 (0.19)	m	t / s.o.	11.5	267 (37)
1000	--	0.36 (0.19)	0.31 (0.24)	m	t / s.o.	6.1	272 (61)
1050	--	0.78 (0.65)	0.72 (0.17)	s.o. / m	t / s.o.	9.0	290 (62)
1100	--	0.74 (0.31)	0.47 (0.21)	m	t / s.o.	13.0	271 (22)
1150	--	0.62 (0.19)	0.74 (0.24)	m	t / s.o.	9.0	306 (20)

KEY; m.minor > s.a.small amount > t.trace   s.d. = standard deviation

Material No.2

TEMPERATURE /degrees C	ENVIRONMENT	WEIGHT CHANGE /% (s.d.)	VOLUME CHANGE /% (s.d.)	XRD RESULTS cristobalite		FTIR RESULTS /% cristobalite	CRUSH STRENGTH /Mpa (s.d.)
				low	high		
UNUSED	--	--	--	--	--	0	222 (37)
800	AIR	1.01 (0.08)	0.32 (0.06)	?	s.t.	0.8	234 (52)
850	--	1.56 (0.20)	0.29 (0.11)	?	t./ s.o.	0.8	265 (29)
900	--	2.55 (0.07)	0.39 (0.23)	s.o.	s.o.	8.8	285 (64)
950	--	2.11 (0.68)	0.76 (0.24)	s.o.	s.o.	2-3	294 (49)
1000	--	1.93 (0.07)	0.30 (0.18)	t./ s.o.	s.o.	2.5	298 (20)
1050	--	2.07 (0.08)	0.89 (0.09)	t./ s.o.	s.o.	4.4	333 (30)
1100	--	1.82 (0.13)	1.40 (0.24)	s.o.	s.o.	2.7	325 (33)
1150	--	1.59 (0.17)	void	t./ s.o.	s.t.	2.5	261 (40)
1200	--	1.76 (0.31)	void	t.	t.	2.5	273 (19)
900	combustion products	2.63 (0.20)	2.06 (0.23)	?	s.o.	1.0	249 (29)
950	--	2.38 (0.15)	0.93 (0.23)	t.	s.o./ m	2-4	314 (46)
1000	--	3.28 (0.06)	0.84 (0.28)	s.o.	s.o./ m	3.4	301 (41)
1050	--	2.27 (0.15)	1.02 (0.32)	s.o.	t./ s.o.	5.0	263 (27)
1100	--	2.51 (0.12)	0.90 (0.28)	s.o.	t	4.2	276 (45)
1150	--	2.28 (0.13)	void	s.o./ m	m	6.3	322 (26)

TABLE 2 LABORATORY OXIDATION TESTS

Material No. 1   : Standard Radiant (outer) Tubes

TEMPERATURE ZONE / degrees C	SERVICE LIFE / months	GLAZE	FTIR RESULTS / % SiO2 (cristobalite)	CRUSH STRENGTH / MPa (s.d.)
600	10	-----	0	147 (55)
960	2	Zircon	18	150 (34)
960	2	Zircon	11	
960	14	-----	16.1	115 (42)
1055	4	Zircon	7	227 (49)
1055	2	-----	17	167 (21)
1055	10	-----	8	214 (34)
1055	14	-----	11.7	299 (38)
1055	2	-----	16	
1070	10	-----	12	103 (15)
1070	10	-----	14	149 (22)
1070	14	-----	7.3	238 (48)
1070	5	-----	7.9	189 (28)
1070	14	-----	10.7	170 (29)
1070	14	-----	7.3	246 (29)
1070	14	-----	10.9	

Material No. 1   : Standard Recirculation (inner) Tubes

700	8	-----	29	86 (19)
960	10	-----	19	78 (19)
960	8	-----	44	74 (19)
960	8	-----	24	225 (28)
1055	4	-----	26	204 (35)
1055	2	Zircon	25	156 (36)
1055	8	-----	40	
1070	2	-----	14	
1070	8	-----	47	182 (24)
1070	8	-----	28	170 (31)
1070	6	-----	15.7	256 (21)
1070	6	TR1301	2.3	234 (41)

Material No. 2   : Standard Recirculation (inner) Tubes

1055	2	yes	13	
1055	6	Zircon	9.8	38 (5)
1070	14	-----	11.5	278 (42)
1070	6	-----	19.5	208 (26)

s.d. — standard deviation

TABLE 3 RADIANT TUBE EVALUATION

Development of ceramic immersion tube for metal melting furnace (carbon-ceramic composite tube)

S Dohi*, H Yokoyama* and T Hosaka**

 * Osaka Gas Co., Ltd., 4-1-2 Hiranomachi, Chuo, Osaka, Japan 541
** Nippon Crucible Co., Ltd., 1-21-3 Ebisu, Shibuya, Tokyo, Japan 150

ABSTRACT: Ceramic immersion tubes for a metal melting furnace, which are widely increasing at present, are inferior in durability and productivity.   To solve these problems, we have developed carbon-ceramic composite heater tube.   Various tests reveal that the new composite tube provides stable and long durability and is economical.

1. INTRODUCTION

To meet the increasing demand for a variety of high quality products, aluminum die cast factories and zinc galvanization factories are increasingly employing, as a melting and holding furnace connected to a die cast machine or an independent holding furnace, an immersion heater furnace with a ceramic tube heater as a heating source because of its small energy consumption and capability of producing molten metal of consistent quality.   However, the average life of the currently prevailing ceramic immersion heater tubes is only about two to three months, according to field records.   Endurance must be improved to enhance productivity.

We began by investigating carbon as a prospective material for immersion heater tubes, in consideration of its various advantageous characteristics:

1) It has very low wettability and does not react with various molten metals.
2) It can endure temperatures of 3,500°C or higher.
3) It provides high thermal and electric conductivity.
4) It is highly resistant to destruction by quick heating or quenching.

Utilizing the above advantages in solving the conventional oxidation problem, we have successfully developed a new carbon-ceramic composite tube that provides long service life and is suitable for combustion-type immersion furnaces.   This report describes the process and results of our development.

2. BACKGROUND OF DEVELOPMENT AND PROBLEMS

2.1 Gas Combustion Immersion Type Metal Melting Furnace

The immersion heating furnace has the following advantages:

① Since molten metal is heated from the inside, thermal efficiency is higher than that of a crucible or reverberatory furnace.

② Compared with a reverberatory furnace, the immersion heating furnace
  • allows deeper molten metal bath and therefore requires smaller installation space,
  • produces less oxide on the molten metal surface due to lower atmospheric temperature in the upper part of the molten metal surface, and
  • heats molten metal from inside so that the molten metal is circulated and stirred by convection, preventing impurities from remaining in the furnace. As a result, the furnace needs less frequent cleaning.

Fig. 1 shows the construction of a gas combustion-type immersion heater, and Fig. 2 the thermal efficiency of the immersion heater.

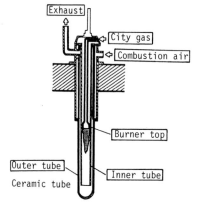

Fig. 1 A Gas Combustion-Type Immersion Heater Tube

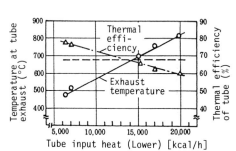

Fig. 2 Thermal Efficiency of Combustion-Type Immersion Heater Tube (for molten aluminum at 680°C)

2.2 Conventional Heater Tube

Various kinds of ceramics are used in conventional heater tubes, most of which are heat-resistant non-oxide ceramics. 'Stala and Hindman (1989) had evaluated and screened this kind of ceramics for an immersion heater tube installed in an aluminum melting furnace.' Typical materials are listed in Table 1.

Table 1 Typical Heater Tube Materials Available on the Market

Class		Material Composition		Manufacturing Method (Sintering Method)
Classification	Symbol	Aggregate Material	Bonding Material	
Refractory grade	A	Silicon carbide	Silicon nitride bonding	Reaction sintering (Nitride sintering)
	B	Silicon carbide + Silicon nitride	Ceramic bonding	Liquid phase sintering
	C	Silicon carbide + Silicon nitride	Carbon bonding	Carbide sintering
	D	Silicon carbide + Graphite	Carbon bonding	Carbide sintering
Advanced ceramics grade	E	Silicon nitride	Silicon nitride bonding	Self sintering
	F	Silicon carbide	Silicon carbide bonding	Self sintering (or reaction sintering)
	G	Sialon	Sialon bonding	Self sintering (or reaction sintering)

① Refractory grade materials (the reaction-sintered product such as silicon nitride-bonded silicon carbide)
Available at a relatively low price. However, with their relatively short life, they do not promise endurance stability in schedule-based operation.

② Advanced ceramic grade materials (such as Sialon self-sintered products)
Compared with refractory grade materials, improvement is expected. However, since their service life varies largely in actual operation, users cannot rely on these materials. Their most disadvantageous feature is their high cost.

These materials were originally developed as materials for furnaces and their parts in contact with molten aluminum or zinc, that is, thermocouple protective tubes, melting furnaces bath, low pressure-casting stokers, graphite crucibles etc., and have come to be applied to heater tubes because of their good achievement record. Therefore, although they are made of heat-resistant, non-oxide compounds of minimum wettability to suppress reaction with molten metal, they do not satisfy all the functions required of heater tube materials.

3. OBJECTIVE AND PROCESS OF DEVELOPMENT

In developing a new heater tube, economy, (namely, low manufacturing cost) is strongly demanded by users and thus is the most important prerequisite to be met. From this standpoint, we analyzed technically the functions and quality that need be satisfied by a new heater tube to ensure long and stable operational durability. Results of this analysis are shown in Table 2.

Table 2   Required Heater Tube Functions and Qualities

Function	Quality
① Heater is isolated from molten metal. (To prevent reaction with molten metal)	→ Corrosion resistance (Impenetrability)
② Resistant to heat generated by heater.	→ Heat resistance (Oxidation resistance)
③ Heat generated by heater is transferred readily to molten metal.	→ High thermal conductivity
④ Resistant to temperature fluctuation by heating or quenching of heater and molten metal.	→ Thermal, structural and mechanical spall resistance

Development was conducted aiming at realization of a heater tube meeting all of the quality requirements. It was necessary to approach the problem from two aspects: 1) to investigate the optimum combination of various heat-resistant non-oxides from the material design aspect, and 2) to study the method of producing and controlling the quality of sintered material of uniform composition from the process design aspect. This section describes the material design technique mainly for molten aluminum and molten zinc.

3.1 Corrosion Resistance (Impenetrability)

When molten metal penetrates the base material of a heater tube, tube quality is degraded. As the heater tube is heated or cooled during operation, the degraded part will crack and eventually break due to thermal stress (structural spalling). Meanwhile, molten metal deposit (the grown deposit of α-Al$_2$O$_3$) grows at the interface between the molten metal and the heater tube, hampering operation efficiency. Accordingly, corrosion resistance is an important quality factor closely related to endurance stability that governs long-term durability of a heater tube. 'The author (1987) showed the design technique used to improve corrosion resistance in Table 3 and (1989) investigated the wettability of Alon to molten pig iron.' In developing the new heater tube, oxides or oxide-nitrides (Alon) were not used, but rather methods 1 through 3 listed in Table 3 were used to ensure endurance stability and long-term durability.

Table 3   Classification of the Methods for Lowering the Reaction between Refractory and Molten Metal

	Principle	Method	Technique
1	Make the mineralogical composition of refractory non-wetting so not to permit the penetration of molten metal	Introducing non-oxide mineral $\Big\{$ Carbon / Carbide / Nitride / Boride	Graphite / SiC / Si$_3$N$_4$, Si$_2$ON$_2$ / B$_4$C, BN
		Introducing new oxide mineral	9Al$_2$O$_3\cdot$2B$_2$O$_3$
2	Make the chemical composition of refractory non-susceptible so not to react with molten metal	Decreasing the amount of reactive component	Less SiO$_2$ / Less Na$_2$O, K$_2$O
3	Make the structure of refractory dense so not to permit the penetration of molten metal	Densification of refractory body	Low porosity / Small pore size / Low permeability

3.2 Heat Resistance (Oxidation Resistance)

To satisfy all of the quality requirements in terms of corrosion resistance, high thermal conductivity and spalling resistance, it is desirable to select heat-resistant, non-oxide materials with minimum oxide content rather than materials from the oxide group.

Generally, the former materials provide high heat resistance and will not melt at the temperature of molten aluminum (about 750°C at maximum) and molten zinc (about 500°C at maximum), or of the heater tube interior (about 1,200°C at maximum). Laboratory tests were conducted on various heat-resistant, non-oxide materials and revealed that carbon with grown crystals is superior to others in all aspects, except for one disadvantage in that it degrades through oxidation. Thus, carbon with grown crystals was selected as one of essential material components of the heater tube.

Nonetheless, carbon and other heat-resistant, non-oxide materials have a disadvantage: they will be oxidized. In other words, carbon is lost in the presence of oxygen such that silicon carbide (SiC) or silicon nitride (Si$_3$N$_4$) is transformed to silica (SiO$_2$), thus eventually leading to deterioration in corrosion and spalling resistances, which are required qualities for a heater tube, and shortening service life. Therefore, in designing a heat-resistant, non-oxide material or a composite of such material, it is necessary to keep in mind that heat resistance is virtually proportional to oxidation resistance.

Accordingly, oxidation suppression of carbon was vital in our development. We investigated various oxidation suppression methods and combined some of them, taking care so as not to deteriorate corrosion resistance. 'The author (1989) investigated the oxidation suppression methods listed in Table 4.'

Table 4 Oxidation Suppression Methods for Carbon Material

	Method		Example
Raw material stage	Selection of raw material	→	Graphite difficult to oxidize is selected.
	Treatment of raw material	→	Oxidation is suppressed by causing P_2O_5 to be adsorbed in the raw material.
Manufacturing stage	Structural reinforcement	Addition of anti-oxidation agent →	Component with low melting point (such as borosilicate glass) is added.
			Metal component (such as Si or Al) is added.
		Partial replacement with material similar to carbon →	Carbon is replaced with carbide (such as SiC or B_4C).
			Carbon is replaced with nitride (such as Si_3N_4 or BN).
	Increase in structural fineness	High pressure forming →	Vacuum forming
			CIP forming
		Impregnation →	Organic substance (such as pitch or resin) is impregnated.
			Inorganic substance (such as phosphate) is impregnated.
	Surface treatment	Coating →	Component with low melting point is applied on the surface.
			Carbide or nitride is applied on the surface.
		Spray coating →	Carbide or nitride is sprayed for coating.

3.3 High Thermal Conductivity

High thermal conductivity which affects energy efficiency in operation is another important quality requirement. Thermal stress generated inside the material can be reduced by increasing thermal conductivity.

Thermal conductivity can be enhanced by 1) increasing structural fineness or 2) selecting a material with high thermal conductivity. The former method is less effective and cannot achieve a substantial increase in the thermal conductivity. In addition, increase in structural fineness brings about another problem: destruction by thermal spalling. Accordingly, we employed the latter method. Specifically, we selected carbon with grown crystals which provides high thermal conductivity and, as material composition, employed a heat-resistant, non-oxide material, specifically carbides (silicon carbide or others) whose thermal conductivity is lower than that of carbon but is still relatively high.

3.4 Spalling Resistance

Spalling resistance is an important quality factor that has significant influence on the endurance stability of a heater tube.

Many of the troubles occurring during operation or particularly in the early stages of operation of a heater tube are caused by structural destruction by thermal stress generated during heating or quenching (spalling destruction). That is, the heater tube is destroyed by one of the following:

1) Mechanical spalling ... Inadequate heater tube shape with respect to the construction of the connecting part causes mechanical stress to concentrate on a local area, resulting in heater tube destruction.

2) Structural spalling ... When the structure of a heater tube is not uniform, the heater tube may be cracked at the interface between different structures, or molten metal may penetrate the structure, forming a deteriorated layer at the boundary, at which the heater tube can crack.

3) Thermal spalling ... The heater tube cracks by thermal stress alone.

4) Combination of 1), 2) and 3).

Using an elaborate production process design, a heater tube with precisely the same configuration as envisioned and with uniform structure was manufactured so as to improve mechanical and structural spalling resistances. Further, material design which prevents cracks from propagating was employed for the corrosion-resistant material to improve structural and thermal spalling resistances.

4. DEVELOPMENT RESULTS

The preceding section has described the quality design method for ensuring endurance stability and long-term durability of a heater tube. This section describes the features and performance of the heater tube developed.

4.1 Features of the Carbon-ceramic Composite Heater Tube Developed

To meet the quality requirements of an immersion heater tube, the developed heater tube has the following features:

① It contains carbon with grown crystals.

② It uses heat-resistant, non-oxide material with minimum oxide component content as the base. More specifically, the material is composed of a non-oxide carbon-carbide (silicon carbide or others) composite.

③ It has a composite bonding structure with high corrosion resistance realized by a carbon bonding sintering method, and high oxidation resistance realized by a reaction sintering method that produces a heat-resistant non-oxide.

④ To suppress oxidation of carbon and the heat-resistant non-oxide, various corrosion-proof, heat-resistant coats are applied depending on the function on the inner and outer surfaces of the heater tube, thus forming a multi-layered structure.

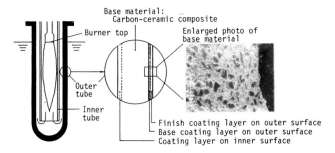

Base material:
Carbon-ceramic composite

Burner top

Enlarged photo of base material

Outer tube

Inner tube

Finish coating layer on outer surface
Base coating layer on outer surface
Coating layer on inner surface

Fig. 3 Construction of Developed Heater Tube

Table 5 Major Characteristics of Developed Heater Tube

Chemical component	C (Graphite)	15%
	SiC	75
	SiO_2	1
Physical property	Porosity	21.5%
	Bulk specific gravity	2.20
	Bending strength	230 kg/cm² (room temperature) 270 kg/cm² (1,200°C)
	Coefficient of linear expansion	3.7×10^{-6}
	Thermal conductivity	20.0 kcal/mh°C
	Mean pore diameter	0.2 micron
Corrosion resistance against molten metal	Aluminum 1,000°C × 450 h Zinc 500°C × 450 h No corrosion	

4.2 Laboratory Test Results

To ensure endurance stability of a heater tube, it is essential to maintain the designed superior quality for an extended period. Tests were conducted in the laboratory on the maintenance of quality of the developed carbon-ceramic composite. The results are as follows.

① Heat resistance (Oxidation resistance)
   Oxidation characteristic of the carbon-ceramic composite (change in weight of the composite exposed to the atmosphere at 750°C and at 1,200°C for 500 hours) is shown in Fig. 4. Change in strength by such heat treatment is shown in Fig. 5. According to these results, there is no weight loss by oxidation, and reduction of strength is minor, indicating that the designated quality is maintained for a long time without deterioration.

② Corrosion resistance against molten metals
   Accelerated corrosion test was conducted at the severe temperature of 1,000°C for 450 hours, using molten aluminum and 500°C for 450 hours, using molten zinc. The test results are shown in Fig. 6 and Fig. 7. Visual observation of a section of the developed composite did not reveal even a trace of corrosion. Neither aluminum penetration nor zinc penetration was detected even under microscopic analysis (EPMA),

as shown in Fig. 6 and Fig. 7. This demonstrates that the developed carbon-ceramic composite provides high corrosion resistance and maintains its designated quality for a long time.

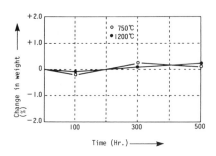

Fig. 4   Oxidation Characteristic (Change in Weight) in the Atmosphere

Fig. 5   Change in Strength

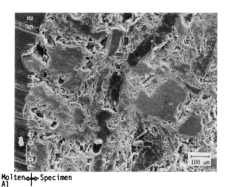

a) Secondary electron image

b) Image of Al element

Fig. 6   EPMA Results of a Specimen after Corrosion Test (1,000°C for 450 hrs with molten aluminum)

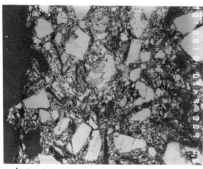

a) Secondary electron image

b) Image of Zn element

Fig. 7   EPMA Results of a Specimen after Corrosion Test (500°C for 450 hrs with molten zinc)

4.3 Field Test of Actual Product

In developing the new heater tube, we conducted a long-term field verification test in addition to the above-mentioned evaluation in long-term laboratory tests.

① Test procedure
Fig. 8 shows an example of a test holding furnace. The developed heater tubes were mounted in a city gas fired holding molten aluminum furnace for an 800 ton die cast machine (holding capacity: 1,000 kg), and gas combustion test was conducted in the furnace without molten aluminum. After investigation of the influence of quick heating and quenching, the furnace was operated under actual operating conditions. The heater tube was taken out to check for damage every month. Thus, we continued the test while confirming the endurance stability of the heater tube.

	Holding furnace for 800 ton die cast machine
Holding capacity	1,000 kg
Molten temperature setting	680°C
Molten metal charging method	Hot charge from central melting furnace
Immersion tube size	ø126 × 550 L (Immersed length: 450 L)
Burner	Self exhaust heat recovery type double tube burner × 3 Combustion capacity: Max 20,000 kcal/hr/burner Min 4,000 kcal/hr/burner Flame is monitored through UV phototube.

Fig. 8 A Gas Combustion Immersion Type Al Holding Furnace

② Test results
The section of the developed heater tube observed after six months of service is shown in Fig. 9. At a glance, the coating film on the inner surface of the heater tube (on the burner side) was partially vitrified, but not to an extent that poses any problem in operation. Neither permeation of aluminum nor melting damage by aluminum were observed on the outer surface in contact with molten aluminum. Fig. 10 shows the results of microscopic analysis (EPMA) on the essential part of the outer surface of the heater tube. Examination shows that the tube is free from aluminum permeation and maintains structural integrity.

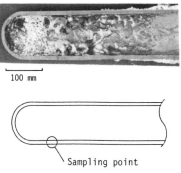

100 mm

Sampling point

Fig. 9 A Section of a Heater Tube after Six Months of Service

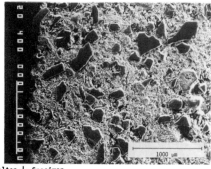 

Molten ←⊦→ Specimen
Al

a) Secondary electron image   b) Image of Al Element

Fig. 10 EPMA Results: A Heater Tube after Six Months of Service

5. CONCLUSION

A number of tests have been conducted on the newly developed carbon-ceramic composite heater tube in a molten aluminum or zinc holding furnace (furnace in actual operation). At present it has been in operation over 12 months in an aluminum holding furnace and over 9 months in a zinc galvanization furnace. The endurance stability and long service life of the composite heater tube are highly appreciated by users on the basis of the following findings in the tests:

① Free from troubles (cracks) which tend to occur in the early stages of operation.
② Molten metal hardly sticks to the surface (high corrosion resistance).
③ Endures quick heating.
④ High thermal efficiency.
⑤ Long operational stability.

Such high acclaim from users justifies the quality design guidelines employed in the heater tube development process, that is, improvement in corrosion resistance, spalling resistance, heat resistance and thermal conductivity.

In addition, the new immersion heater tube has an economic advantage over its conventional counterparts; manufacturing cost is 4/5 or less that of the commercially available refractory grade products and 1/3 to 1/10 or less that of advanced ceramics grade products.

We plan to continue our efforts in promoting the use of the new immersion heater tubes and gas combustion immersion type metal melting furnaces and development of a new type heating furnace.

REFERENCE

Hosaka T et al. 1987 Taikabutsu 39(7) pp2-5
Hosaka T et al. 1989 Ceramic Engineering Handbook pp1663
Stala C and Hindman DL 1989 Prep. International Gas Research Conference
    III pp186-195
Takeda K and Hosaka T 1989 Interceram 38(1) pp18-22

Thermomechanical performance of ceramic radiant tubes

J. R. Hellmann and A. E. Segall

Center for Advanced Materials, The Pennsylvania State University, University Park, PA 16802

ABSTRACT: A test and design methodology for predicting thermomechanical performance of gas-fired ceramic radiant tubes is under development within the Center for Advanced Materials with funding from the Gas Research Institute.

The technical approach involves modeling of thermal profiles and the related thermoelastic stress distributions present in the tubes, during transient and steady state heating, followed by prediction of thermomechanical failure using Weibull probabilistic analysis and measured mechanical properties of the tube. Results of the analyses indicated that the failure probability of the ceramic tube under transient and steady state conditions was less than 16/1000 (0.016) for a simply supported tube without radial or axial constraints.

1. INTRODUCTION

Radiant tubes are used in many industrial combustion-heated processes that require precise control of the furnace ambient and exclusion of the combustion gases from the workload. In applications such as melting, sintering, and heat treating, heat is transferred to the workload by radiation from the tube, which is internally heated by a burner. Until recently, the use of radiant tube technology has been limited to process temperatures below 1100°C due to temperature limitations of the various metal alloys from which conventional radiant tubes are constructed; however, newer materials, such as nonoxide monolithic and composite ceramics, are now available in the sizes required for such high temperature structural applications as radiant tubes.

The advantages of substituting ceramics for the alloy materials in these applications include higher process temperatures and heat fluxes than are possible with alloys, in combination with improved corrosion resistance in aggressive environments (Liang and Schreiner 1986). A comparison of the anticipated performance of ceramic radiant tube systems versus conventional alloy radiant tube systems in industrial processes is given in Table 1.

Although the advantages of ceramics for high temperature applications have long been recognized, their incorporation as structural components in high temperature systems has been delayed as a result of early, marginally successful attempts to use these materials in equipment designed to accommodate the more forgiving thermal and mechanical characteristics of metals. The successful application of ceramics as high temperature structural components requires the development and verification of a new design methodology that can be used to manage the characteristic brittleness and thermal shock susceptibility of these materials. Such a design methodology would reflect the difference in thermophysical, thermomechanical, and thermochemical properties of ceramics as compared to metals, and would remove current barriers to the full exploitation of their potential as high temperature structural components.

Table 1. State-of-the-art versus emerging technology in indirect gas-fired heating systems (after Liang and Schreiner, 1986)

	State of the art	Goal
Material	Metal	Ceramic
Furnace Temperature, °C	760-1,038 (Max.)	1,038-1,427
Furnace Efficiency (Recuperated), %	40-65	70-85
Heat Flux, W/m^2	19,000-25,000 (Max.)	63,000-95,000
Tube Life, Year	0.5-3	Minimum 2

Table 2. Properties of SCRB210 and Hexoloy SA at room temperature

Property	SCRB210	Hexoloy SA
Youngs Modulus (E)	366 GPa	410 GPa
Poisson's ratio (v)	0.16	0.14
Shear modulus (G)	158 GPa	180 GPa
Coefficient of thermal expansion (α) (room temperature to 1000°C)	4.6 x 10^{-6}/°C	4.02 x 10^{-6}/°C
Density (ρ)	3.1 gm/cm^3	3.1 gm/cm^2
Thermal Conductivity (k)	97.7 W/m-°C	72.6 W/m-°C
Thermal Diffusivity (D)	0.846 cm^2/sec	0.842 cm^2/sec
Area characteristic strength (σ_0)	119 MPa	167 MPa
Weibull modulus (m)	7.2	15.2

Our approach to the development of a radiant tube test and design methodology was to assess the state-of-the-art in ceramic radiant tube technology, to improve our predictive capabilities, and to experimentally verify the thermal and mechanical performance of ceramic radiant tubes in simulated industrial applications. A review of conventional radiant tube technology by Harder, et al. (1987) revealed that although a wealth of data was available for modeling the performance of alloy radiant tubes, insufficient data were available to accurately and reliably model the thermal and mechanical performance of ceramic materials currently proposed for radiant tube applications. A more thorough knowledge of temperature distributions in ceramic radiant tubes was required to calculate thermoelastic stress distributions and tube failure probabilities during transient and steady state operation. This paper deals with the experimental quantification of transient and steady-state axial and circumferential profiles in silicon-carbide radiant tubes and their influence on the resulting thermoelastic stresses and Weibull based failure probabilities predicted using finite-element analysis. Results indicated that for the materials, tube geometries, and furnace configurations studied, the resulting transient and steady-state axial and radial temperature gradients are not severe enough to induce thermoelastic stresses sufficient to induce high failure rates.

2. RESULTS

The ceramics evaluated in this study were: 1) a reaction bonded silicon carbide material[1] (SCRB210), and; 2) a sintered alpha silicon carbide material[2] (Hexoloy SA) that were specifically developed for such high temperature structural applications as heat exchangers, burner nozzles, and radiant tubes. The SCRB210 material is a nearly fully dense ceramic assemblage of bimodally distributed silicon carbide grains (with "mean" grain sizes of 4.5 and 55 μm) with approximately 15 volume percent of the ceramic being comprised of residual silicon at the grain boundaries (Figure 1). The Hexoloy SA is a nearly fully dense polycrystalline form of alpha silicon carbide with an average grain size of 5.0 μm (Figure 2); boron and carbon have been added as sintering aids. The materials' high thermal conductivity rivals or exceeds that of the majority of the high temperature alloys available for structural applications (Metals Handbook 1980) and, when taken together with their low thermal expansion coefficient, low density, and moderate strength (Table 2), contributes to the materials' excellent thermal shock resistance relative to other structural ceramics. The combination of high thermal conductivity, thermal shock resistance, and demonstrated corrosion resistance in high temperature oxidative environments (Stillwagon, et al. 1988) makes silicon carbide an excellent candidate for radiant tube applications.

For this study, 1.9 m long silicon carbide tubes were tested in a variable firing rate gas combustion furnace test cell at Columbia Gas Systems Services Corp. (CGSSC); both 10 cm diameter and 20 cm diameter x 0.95 cm thick wall tubes of SCRB210 and 10 cm diameter x 0.64 cm thick wall tubes of Hexoloy SA were evaluated. The tubes were simply supported within the test cell to permit radial and axial thermal expansion to occur unopposed (Figure 3). Tube instrumentation consisted of remote optical pyrometers and thermocouples mounted on the external tube wall to characterize axial and circumferential temperature gradients imposed by variations in combustion flame temperature. Rapid response thin foil type K thermocouples (chromel-alumel) were used to characterize the transient thermal profiles, whereas type S (Pt-Pt10Rh) thermocouples were employed at the higher temperatures characteristic of steady-state tube operation. Thermocouple placement was based on assessments conducted at CGSSC using transparent quartz tubes to determine flame length and on preliminary measurements of spatial temperature variation for the 100-230KW firing rates used during testing. To secure the thermocouples and to avoid degradation from contact with the silicon carbide, a high temperature inorganic cement[3] was used to form a passivating layer between the tube and thermocouple bead. Silicon carbide fibers[4] were also used to strap each thermocouple to the

---

[1] Coor's Ceramics, Golden, Colorado

[2] Carborundum Co. Structural Ceramics Division, Niagara Falls, New York

[3] Sauereisen No. 8, Sauereisen Cements Co., Pittsburgh, PA

[4] Nicalon fibers, Nippon Carbon Co. Ltd., Tokyo

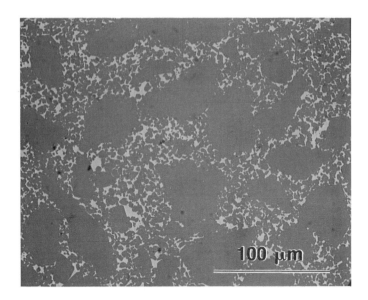

Figure 1.   Microstructure of the SCRB210 reaction bonded silicon carbide illustrating the bimodal grain size distribution and residual silicon in the grain boundaries

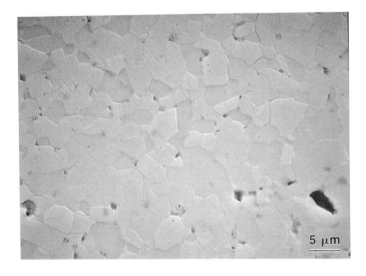

Figure 2.   Microstructures of Hexoloy SA sintered alpha silicon carbide

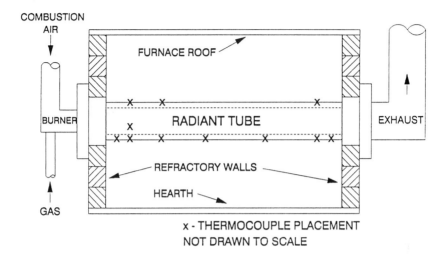

Figure 3.   Schematic of radiant tube test cell and thermocouple placement

Figure 4.   Radiant tube with externally mounted thermocouples visible ($T_{avg.}$ = 1260°C, firing rate = 230 KW)

tube's outer surface to assure intimate mechanical contact in the event of bond failure. Figure 4 shows an SCRB210 radiant tube during steady-state operation with the cemented thermocouples and straps clearly visible.

The furnace was fired at the prescribed rate from ambient temperature. Location-dependent temperature data was recorded continuously from startup with a high speed computer up to an average tube temperature of 1260°C. Data acquisition intervals were set at 1 second in order to capture the transient as well as steady-state profiles shown for the maximum firing rate of 230KW in Figure 5. Subsequent tests at lower firing rates yielded no significant variations in the shape of the transient heat-up curves or steady-state axial profiles. The similarities in axial temperature profiles under variable firing rates are most likely the result of the dependence of the flame temperatures on burner design, tube size and material, and the tendency of the system efficiency to decrease with higher firing rates.

2.1 Thermal Profile Analysis

In order to model the tubes' transient response, axial and circumferential temperatures were measured using low profile, rapid response thermocouples mounted to the outer surface of the tube. Attempts at mounting the thermocouples to the internal surface yielded measurement of excessively high temperatures because the relatively high cement profile caused the thermocouple to sense the boundary layer and flame temperatures. This problem was overcome by using the measured external temperatures in conjunction with heat exchange equations and finite-element analysis to model the tubes' transient response. In this analysis, a quasi-static external heat transfer coefficient based on the superpositioning of convective and radiative heat transport was calculated. Although natural convection was not expected to contribute greatly to the heat removed from the tube, as a conservative measure it was included in the calculations with expressions developed for horizontal cylinders (Rohsenow, et al. 1985). Estimates of radiative heat transport were based on a network analysis (Holman 1976) that considered the energy emitted and absorbed by the tube and surrounding furnace surfaces as a function of time and measured tube and furnace temperatures. The resulting radiation exchange equations:

$$ J_i = \frac{1}{1 - F_{ii}\left(1 - \varepsilon_i\right)} \left[\left(1 - \varepsilon_i\right) \sum_{j \neq i} F_{ij} \, J_j + \varepsilon_i \, \sigma T_i^4 \right] \tag{1} $$

$$ q_i = \frac{\varepsilon_i}{1 - \varepsilon_i} \left[\sigma T_i^4 - J_i \right] \tag{2} $$

relate the Stefan-Boltzmann constant, σ, tube and furnace emissivities, ε_i, temperatures, T_i, radiation view factors, F_{ij}, radiosity, J_i, and heat flux, q_i. The subscripts i and j allow the tube, furnace, and any thermal load to be divided into as many uniform temperature segments as desired. A furnace emissivity of 0.90 was used based on the insulation manufacturers product data. A tube emissivity of 0.79 was selected on the basis of a comparison of tube surface temperatures measured using direct contact thermocouples and optical pyrometry. Once the radiative flux was determined by simultaneously solving equations 1 and 2, a radiative coefficient was calculated by dividing the flux by the temperature difference between the tube wall and furnace environment. Figure 6 illustrates the strong temperature dependency of the resulting heat transfer coefficient for both the 10-cm and 20-cm tubes and the relatively minor contribution of natural convection. It is important to note, however, that the values shown in Figure 6 are conservatively based on the maximum tube temperature measured at each time interval.

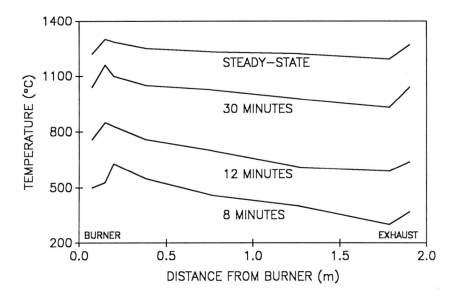

Figure 5. Transient and steady state axial temperature profiles (20 cm diameter SCRB210)

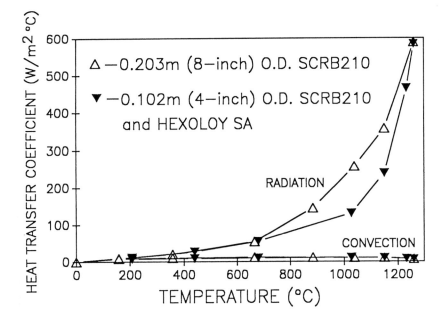

Figure 6. Temperature dependence of the radiative and convective heat transfer coefficient for 10-cm and 20-cm radiant tube tests

The nonlinear heat transfer coefficient calculated by the thermal network equations was used in the ANSYS (DeSalvo and Swanson 1975) finite-element package with the materials' temperature-dependent thermal properties and the measured external temperatures to model the thermal transients experienced by the ceramic radiant tubes during cold starts. Cold start thermal transients were modeled by iteratively imposing temperature loads on the internal surface until the external temperatures matched the experimentally measured values. Solution of the differential equations governing the two-dimensional diffusion of heat through the material and the release of this heat to the furnace environment (nonlinear heat transfer coefficient) was handled by ANSYS. This process was repeated in a piece-wise linear fashion at selected time intervals until the entire transient and the most severe axial and radial gradients were modeled. Because of the initially high diffusivity at the lower starting temperatures, the most severe radial gradients did not occur at the start of firing, as was originally expected. Instead, the combined effects of lowered thermal diffusivity, lagging furnace temperatures, and greater radiative heat transfer at higher temperatures caused the largest radial gradients to occur well into the transient. Figure 7 shows the calculated radial temperature differences for axial positions where the most severe axial gradients occurred (hot spots at positions approximately 20 cm and 46 cm along the 20-cm SCRB210 and 10-cm SCRB210 and Hexoloy SA tubes, respectively). A check of the predicted steady-state thermal flux was within 5% of the estimates at CGSSC for both 10-cm and 20-cm tube geometries. It is important to note, however, that these calculations were dependent on the silicon carbides' ability to conduct and diffuse heat. Consequently, a thorough understanding of the candidate materials' thermal diffusivity over the expected operational temperature range was needed.

Thermal diffusivity was measured at temperatures up to 1400°C for SCRB210 and Hexoloy SA using a state-of-the-art laser flash diffusivity apparatus. Over the temperature range examined, the two materials' thermal behavior was identical (Figure 8); tubes of similar dimensions and exposed to similar firing conditions should experience the same transient thermal gradients. Also of interest was the relatively high thermal diffusivity exhibited by these materials, as shown in Figure 8 by the comparison to a typical alloy (Haynes 188) used in metallic radiant tube. The high thermal diffusivity is responsible for minimizing thermal gradients and the severity of thermoelastic stresses during transient operation.

An analysis of the measured transient and steady-state tube temperature profiles (Figure 5) shows axial gradients resulting from variations in flame temperature and heat transfer coefficients along the tube's length. The maxima on the left portion of the profiles can be attributed to the location of maximum flame temperature. The decreasing trend in the tube's surface temperatures upon approaching the exhaust end of the tube reflects the decaying intensity of the flame temperature with distance from the burner. At the exhaust end, however, the surface temperature rise is believed to be caused by back-radiation from the 90 degree bend in the exhaust flue. The 90 degree bend allows direct combustion flame impingement on the exhaust flue which may yield an additional radiative thermal flux back into the tube. Secondary flows of hot exhaust gases back into the tube may also be a factor because of the flue's large diameter relative to the length in the exhaust plenum where an abrupt change in exhaust direction occurs. Temperatures recorded at circumferential locations along the tube axis indicate no discernable circumferential variation on the outside surface. These findings indicate that under the current testing conditions, only axial and radial transient temperature gradients can be expected. Consequently, the stress analysis must consider the effects of both temperature gradients.

2.2 Thermoelastic Stress Analysis

Under the influence of radial and axial temperature gradients, and neglecting edge effects, the circumferential, axial, and radial thermoelastic stress components can be derived through a modification of the classical thermoelastic relationships (Timoshenko and Goodier 1970):

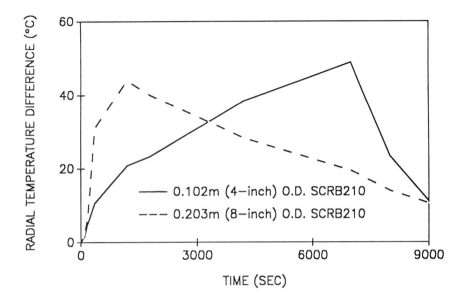

Figure 7. Calculated radial temperature differences at hot spots near the burner for 10-cm and 20-cm SCRB210 tubes

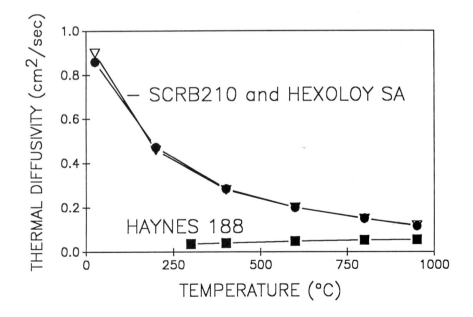

Figure 8. Temperature dependence of thermal diffusivity for Hexoloy SA and SCRB210 silicon carbides relative to a conventional high temperature alloy (Haynes 188)

$$\sigma_\theta = \frac{\alpha E}{(1-\upsilon)r^2}\left[\frac{r^2 + r_i^2}{r_o^2 - r_i^2}\int_{r_i}^{r_o} T(r)r\,dr + \int_{r_i}^{r} T(r)r\,dr - T(r)r^2\right] + \tau_\theta\left(T''(x)\right)$$

$$\sigma_z = \frac{\alpha E}{(1-\upsilon)}\left[\frac{2}{r_o^2 - r_i^2}\int_{r_i}^{r_o} T(r)r\,dr - T(r)\right] + \tau_z\left(T''(x)\right) \qquad (3)$$

$$\sigma_r = \frac{\alpha E}{(1-\upsilon)r^2}\left[\frac{r^2 + r_i^2}{r_o^2 - r_i^2}\int_{r_i}^{r_o} T(r)r\,dr - \int_{r_i}^{r} T(r)r\,dr\right] + \tau_r\left(T''(x)\right)$$

where E is the modulus of elasticity, α the coefficient of thermal expansion, υ is poisson's ratio, r the radius of evaluation, $T(r)$ the radial temperature distribution, and r_o and r_i are the outer and inner radii, respectively. The terms, τ_θ $(T''(x))$, τ_z $(T''(x))$, and τ_r $(T''(x))$ account for the additional stresses induced by the axial temperature gradients; these stresses are directly proportional to the curvature of the axial thermal gradients, $T''(x)$. Although analytical solutions exist for the axial and radial gradient imposed stress state described by Equation 1 (Lee 1966; Borisi and Sidebottom 1985), they are based on infinite cylinder assumptions and do not describe free edge effects. Because the most severe stresses were expected to occur in the vicinity of the free edges, and due to the laborious nature of the stress and subsequent Weibull calculations, finite-element analysis was again used.

A finite-element model consisting of 900 axisymmetric elements (three elements across the tube wall thickness) was used to model the simply supported tube. Mesh adequacy was verified by comparing stresses calculated at locations far from the free ends to values predicted by Equation 3 and to a 5400 element mesh (6 elements across the tube wall thickness). Radial, circumferential (hoop), and axial stresses caused by the imposed axial and radial temperature gradients were calculated using the ANSYS finite element code. Figure 9 illustrates the axial stresses experienced by the 10-cm and 20-cm SCRB210 tubes during the worst transient gradients. The local effects of the axial gradients can be seen on the left side of the plots where the stresses show large changes in magnitude. These changes in magnitude are caused by the localized hot region and the resulting severe axial gradients near the burner end. At both free ends, the stresses rapidly decay to zero to match the stress-free boundary conditions. Although not shown, the circumferential (hoop) stresses are similar in distribution and magnitude except for the free edges where they do not decay. The radial stresses are compressive and at least two orders of magnitude lower than the axial or hoop stresses and, consequently, were not used in subsequent failure probability calculations.

2.3 Prediction of Tube Reliability

Once the surface and volume stresses were calculated, mechanical failure predictions were made using the temperature, surface, and volume dependent Weibull strength data measured on tubular sections of the material. Weibull strength data was measured in diametral compression on O-ring and C-ring specimen geometries (Hellmann and Kennedy 1988); these geometries were employed to evaluate the effects of different flaw populations inherent on the inner and outer surfaces of the tubes, respectively. Figures 10 and 11 show temperature dependent Weibull strength plots with characteristic strengths and Weibull moduli determined from the O-ring and C-ring specimens, respectively. Volumetric tube strengths were conservatively based on the O-ring data because of the lower characteristic strengths and Weibull moduli (more

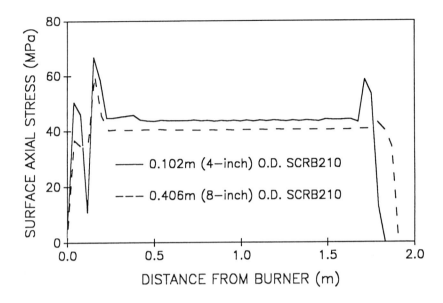

Figure 9.   Axial surface stresses developed in 10-cm and 20-cm reaction-bonded silicon carbide tubes

severe flaw population). Assumptions of independence of both the principal and surface stresses were initially used in evaluating the two parameter Weibull stress volume and stress area integrals (Johnson 1979):

$$P_f = 1 - P_s = 1 - \exp\left\{-\int_V \left(\frac{\sigma}{\sigma_{o_V}}\right)^{m_V} dV - \int_A \left(\frac{\sigma}{\sigma_{o_A}}\right)^{m_A} dA\right\} \qquad (4)$$

where P_f represents the probability of failure, P_s is the probability of survival, m_V and m_A are the Weibull moduli for volume and area (internal and external) respectively, σ is the calculated principal and surface stresses (all tensile components), and σ_{o_V}, σ_{o_A} are the characteristic unit tension strengths for volume and area flaw distributions, respectively. Integration of Equation 4 was performed for each element, the internal and external surfaces, and the volume of the tube. The total failure probability was calculated using the principle of independent action (PIA):

$$\left(1 - P_{f_{PIA}}\right) = (1 - P_{f1})(1 - P_{f2})...(1 - P_{fn}) \qquad (5)$$

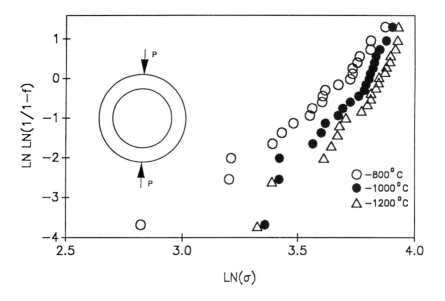

Figure 10. High temperature O-ring Weibull strength distribution (internal surface initiated failure) for SCRB210. Area based Weibull moduli and characteristic strengths for 800°C, 1000°C, and 1200°C are: 4.8, 182 MPa; 7.6, 215 MPa; and 7.3, 229 MPa, respectively

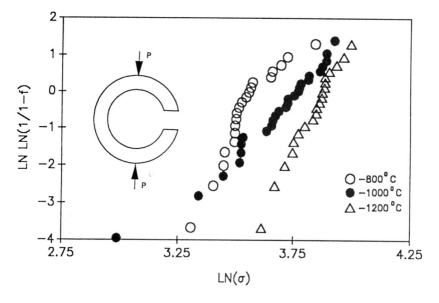

Figure 11. High temperature C-ring based Weibull strength distribution (external surface initiated failure) for SCRB210. Area based Weibull moduli and characteristic strengths for 800°C, 1000°C, and 1200°C are: 9, 220 MPa; 7.2, 251 MPa; and 12.9, 297 MPa, respectively

which assumes that the stress components acted independently without crack interaction (Stanley, et al. 1973). A biaxial stress approximation (Batdorf 1977) was then used to determine an upper bound for the failure probability of each tube. Table 3 lists the results of these calculations for both materials and tube geometries.

The exponential nature of the failure probability calculations are shown by Figure 12. These values are based on the measured axial gradients and calculated radial differences corresponding to steady-state heat-output efficiencies from 0% to 100% of the heat-input. Although the stresses are directly proportional to changes in ΔT_r, the resulting failure probabilities will show sharp increases. Because the maximum stresses (outer surface) are nearly equibiaxial, a biaxial stress approximation was also used to calculate the second curve shown in Figure 12. Even with the biaxial stress approximations, the steady-state failure probability remains low (<0.7%) for the expected heat-output efficiency of ≈25%. It is important to note, however, that these values are also based on the measured axial gradients that are a function of burner and furnace design; each system must be analyzed separately. Because the same system was used in all of the tests, similar results were obtained for the Hexoloy SA material.

The effects of stress-volume and stress-area interaction can be seen by a comparison of the two different SCRB210 tubes. Although the average stress level of the smaller tubes was 11% higher (larger radial and axial gradients) than the 20-cm tubes, their transient failure probabilities do not differ significantly because the Weibull analysis considers the stress magnitudes as well as the surface area and volume exposed to tensile stresses. Although the stresses are larger for the small tubes, the net amount of area and volume exposed to tensile stresses is lower, which underscores a difference in design protocol for ceramic components versus the more traditional alloys. Generally, in alloy components, larger sections are employed to decrease the stress and increase the survivability of the component.

A comparison between the small SCRB210 tube and the Hexoloy SA tube also reveals the significance of the differences in Weibull strength parameters. Although the axial and radial stress gradients are identical, the transient failure probabilities are significantly different because of the less severe (higher characteristic strength) and more uniform (higher Weibull modulus) flaw populations characteristic of the Hexoloy SA material. Nevertheless, the low overall failure probabilities for both materials indicate that, in the absence of slow crack-growth, creep related failures, or mechanical clamping, both materials are viable candidates for radiant tube applications involving cold-start transients and steady-state operation at temperatures approaching 1300°C. Tests using large thermal loads have also shown, that, in the absence of slow crack-growth, both of the materials and tube geometries are capable of surviving significantly more severe transients. Quantification of these results are currently underway.

3. SUMMARY

A test and analysis methodology was developed for calculating thermal profiles and thermoelastic stress distributions, and for predicting the mechanical reliability of silicon carbide radiant tubes during transient and steady-state heating. Experimental studies focussed on two commercially available silicon carbides: 1) a reaction bonded silicon carbide (SCRB210), and; 2) a sintered alpha silicon carbide (Hexoloy SA). Direct tube instrumentation and electronic data acquisition of the tubes' thermal response during testing was combined with thermal flux calculations to model the time-dependent thermal profiles during a cold-start transient and into steady-state heating. Finite element analysis was used to calculate the circumferential, axial, and radial thermoelastic stress distribution from the thermal profile data and the materials' temperature-dependent properties. Results indicate nearly equibiaxial surface stress distributions in the axial and circumferential (hoop) directions; radial stresses were always compressive (for a cold start scenario) and were significantly lower than the axial or hoop stresses. Weibull probabilistic analyses performed using the calculated thermoelastic stress distributions and strength distribution data determined on C-ring and O-ring test geometries indicated that, in the absence of slow crack growth, creep, or mechanical clamping, both materials are viable candidates for radiant tube applications to approximately 1300°C.

Table 3. Predicted tube failure probabilities for the SCRB210 and HEXOLOY SA radiant tubes under transient and steady-state heating

Diameter (m)	Length (m)	Material	Failure Probability (%)	
			Transient	Steady-State
0.203 (8-inch)	1.9 (75-inch)	SCRB210	1.6	0.013
0.102 (4-inch)	1.8 (72-inch)	SCRB210	2.5	0
0.102 (4-inch)	1.8 (72-inch)	Hexoloy SA	0.02	0

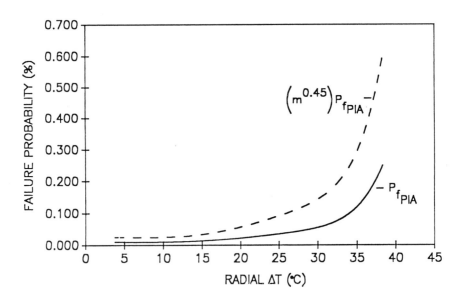

Figure 12. Surface and volume based failure probability predictions as a function of calculated radial temperature differences for 20 cm diameter SCRB210 tubes at 230 KW input

4. ACKNOWLEDGMENTS

Special thanks are are extended to the staff at Columbia Gas Systems Service Corporation for their assistance during testing. This project was funded by the Gas Research Institute under contract number 5084-238-1302.

5. REFERENCES

S. B. Batdorf, "Some Approximate Treatments of Fracture Statistics for Polyaxial Tension," Int. J. of Fracture, Vol. 13, No. 1, February 1977, pp. 5-10.

A. P. Borisi and O. M. Sidebottom, "Advanced Mechanics of Materials," pp. 521-523, John Wiley and Sons, 1985.

G.N. DeSalvo and J.A. Swanson, ANSYS: Engineering Analysis System User's Manual, Swanson Analysis Systems, Inc., Houston, PA 1975.

R.F. Harder, R. Viskanta, and S. Ramadhyani, Gas-Fired Radiant Tubes: A Review of Literature, topical report to Gas Research Institute, GRI-87/0343, December, 1987, available through NTIS.

J. R. Hellmann and B. K. Kennedy, eds. Projects Within the Center for Advanced Materials, annual report to Gas Research Insitute, report no. CAM 8807, GRI 88/0181, June 1988, available through NTIS no. PB89-142624.

J.P. Holman, Heat Transfer, 4th edition, McGraw Hill, New York, 1976, p.333.

C. A. Johnson, "Fracture Statistics in Design and Application," G. E. Report No. 79CRD212, December 1979.

C. W. Lee, "Thermoelastic Stresses in Thick-Walled Cylinders Under Axial Temperature Gradients," J. Appl. Mech., Trans. ASME, 33, 1966, pp. 467-469.

W. W. Liang and M. E. Schreiner, "Advanced Materials Development for Radiant Tube Applications," Proceedings, 1986 Symposium on Industrial Combustion Technology, Chicago, IL, April 1986.

Metals Handbook, Vol. 3: Properties and Selection: Stainless Steels, Tool Materials, and Special Purpose Metals, 9th edition, American Society for Metals (1980).

W.M. Rohsenow, J.P. Hartnett, and E.N. Ganic, editors, p. 6-42 in Handbook of Heat Transfer Fundamentals, 2nd edition, McGraw Hill, New York, 1985.

P. Stanley, H. Fessler, and A. D. Sivill, "An Engineer's Approach to the Prediction of Failure Probability of Brittle Components, " Proc. Brit. Cer. Soc., Vol. 22, No. 3, March 1973, pp. 453-487.

T. L. Stillwagon, D. M. Kotchick, M. McNallan, J. Golemboski, and M. G. Coombs, Performance Verification of Industrial ceramic Materials - Environmental Effects on Performance, topical report to the Gas Research Institute, GRI 88/0144, April 1988, available through NTIS.

S. P. Timoshenko and J. N. Goodier, "Theory of Elasticity," McGraw-Hill, New York, 1970, pp. 448-449.

High temperature liquid corrosion of silicon nitride based materials

G. Gauthier, M.A. Lamkin, F.L. Riley and R.J. Fordham[*]

Division of Ceramics, School of Materials, The University of Leeds,
Leeds LS2 9JT.

[*]Commission of the European Communities, Joint Research Centre,
Petten, ZG 1755, The Netherlands.

ABSTRACT: Silicon nitride has been corroded at high temperature in the
presence of fixed amounts of sodium and vanadium-containing compounds.
A fast oxidative corrosion stage is followed by a slow oxidation
reaction. The corrosion process is modelled in terms of the solution
of the primary oxidation product, silicon dioxide, in a liquid silicate
surface film. The overall processes of corrosion and oxidation are
controlled by phase-equilibrium relationships in the metal oxide-
silicon dioxide system, and by the dissolution of silicon dioxide.

1. INTRODUCTION

Silicon nitride has been widely proclaimed as the high temperature internal
combustion engine material of the future. Realism has now tempered
enthusiasm, but silicon nitride based materials are currently finding use
in Japan on a small scale for automobile engine components and this
activity will certainly provide a base for further developments [1,2].
Less ambitious and more mundane possible applications in other areas are
likely to include small sliding wear parts, hot metal forming die components, and heat and wear resistant components used in the areas of pulverised coal, and gas, combustion, Most of these applications require a
resistance to oxidising conditions, possibly in the presence of metal
oxides, at temperatures up to 1500°C.

Silicon nitride oxidises under normal atmospheric conditions to silicon
dioxide:

$$Si_3N_{4(s)} + 3O_{2(g)} = 3SiO_{2(s)} 2N_{2(g)}; \quad \Delta G^\circ_{1500K} = -1063 \text{ kJ mol}^{-1} \quad ...(1)$$

Oxygen mobility in silicon dioxide is very low and it thus provides an
effective barrier layer against continuing oxidation. The stability of
the silicon dioxide film is consequently of considerable importance for
the life of a component. A major factor determining stability is the
possibility of reaction with metal salts or oxides, to form silicates, in
which oxygen mobility may be greater by many orders of magnitude.
Summarised data [3] for oxygen diffusion coefficients in silicon dioxide
and silicates illustrating differences in oxygen mobility are shown in
Figure 1.

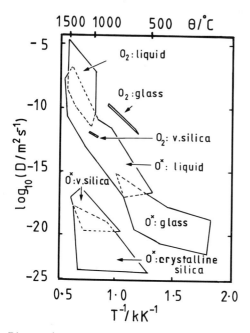

Figure 1: Collected data for oxygen diffusion in silicon dioxide and silicates. "O_2" data refer to diffusion of molecular oxygen, "O^x" data to tracer diffusion measurements. The term "liquid" refers to systems at temperatures above the glass transition temperature, "glass" below.

Two likely sources of metal oxide are:

i)   the silicon nitride intergranular phase, inevitably present because of the necessity for using liquid-forming sintering aids, for example MgO, Al_2O_3 and Y_2O_3, in the production of dense sintered silicon nitride. These additives have been selected on the basis of their efficiency in promoting sintering of silicon nitride powder, and for the resulting materials' mechanical properties;

ii)   vapour phase contaminants, due to the presence of compounds such as sodium sulphate or vanadium pentoxide in combustion air or fuel oil. These can form, at normal operating temperatures, liquid oxide films on the silicon nitride surface and interact with the silicon dioxide to allow fast corrosion. The presence of external contaminants is in practice the more serious because of the larger quantities available. At high temperatures, and in flowing air conditions, salts such as sodium chloride or sulphate can react rapidly to form the corresponding silicates. For example:

$$Na_2SO_{4(1)} + SiO_{2(s)} = Na_2SiO_{3(1)} + SO_{3(g)}; \quad \Delta G^o_{1273 \ K} = 147 \ kJ \ mol^{-1} \quad (2)$$

and provided p_{SO_3} is sufficiently low (< 0.93 μbar at 1273 K) the reaction goes to the right.   The physical solubility of SiO_2 in Na_2SO_4 is, however, very small [4] and under higher pressure of SO_3 when sodium sulphate is stable, no reaction with silicon dioxide occurs [5].

Silicon dioxide is formally an "acidic" oxide, and at high temperature it readily reacts with most metal oxides to form silicates. In complex multi-component systems quite low (< 800°C) eutectic temperatures can be found, and partially liquid reaction products are common at the temperatures of interest (900-1500°C). The stability of a protective silicon dioxide film therefore depends on the system composition, as is illustrated in the schematic binary diagram (Figure 2), where for composition A, to the MO-rich side of the liquidus curve, at temperatures above the eutectic, SiO_2 cannot exist.

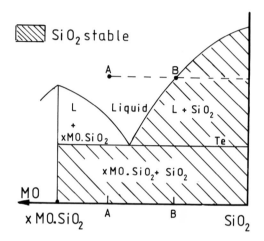

Figure 2: Schematic of a hypothetical SiO_2 binary system, showing stability regions for SiO_2.

Fast oxidative corrosion is in practice then seen. Whether composition B can be attained, to allow the development of a protective silicon di-oxide film, depends on the circumstances:

i)   a fixed quantity ('static' corrosion) of metal oxide capable of reacting with silicon dioxide will normally allow an initial period of fast reaction, in which the silicon dioxide dissolves in the liquid at approximately the same rate as it is formed. Eventually, with continuing production of SiO_2, the system composition shifts to B on the liquidus surface, and a transition to slow oxidation occurs;

ii)   with a continuous supply of metal oxide ('dynamic corrosion'), equilibrium (composition B) cannot be attained if the metal oxide flux exceeds a value given by:

$$j_{MO} = j_{SiO_2} \frac{(1 - x_b)}{x_b} \qquad \qquad \dots \dots (3)$$

where $j_{(SiO_2)}$ is the rate of production of silicon dioxide by normal oxidation and x_b is the mole fraction of SiO_2 at composition B. In practice oxygen permeation through silicon dioxide is so slow that even small metal oxide fluxes (> 10 µbar pressure for example at 1000°C) are sufficient to bring the film composition to point B. Under these circumstances the most favourable case corresponds to a system having a high MO/SiO_2 ratio at the liquidus composition. This is a feature of the

SiO_2-V_2O_5 system, a section of which is shown in Figure 3, after reference [6]:

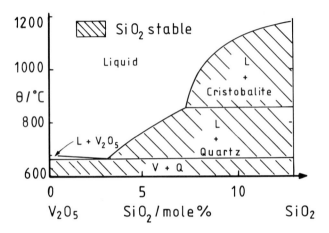

Figure 3: Section of the V_2O_5-SiO_2 system, showing significant solution of SiO_2 in a V_2O_5-rich liquid, and the SiO_2 stability region [6].

In this simple binary system ~ 9 mole % of SiO_2 is soluble in the liquid at 1100°C. Under conditions of 'dynamic' corrosion, because equilibrium leading to the development of a protective silicon dioxide film cannot be attained, complete destruction of the silicon nitride must eventually result.

Practical systems are more complex because of the availability of inter-granular phase cations such as Y^{3+}, which move the liquidus envelope closer to silicon dioxide. This is contrary to the requirement of equation (3), but the actual magnitude of the effect in, for example, the Na_2O-SiO_2-Y_2O_3 system is not marked [7]. For this reason the liquid corrosion behaviour of silicon nitride materials appears not to be greatly affected by the nature or quantity of the densification aids used.

The overall extent of corrosion thus depends on the availability of the corrodant. Thin films of a salt or metal oxide applied to a silicon nitride surface, or a single exposure to corrodent vapour, give fast, approximately parabolic, kinetics suggesting a diffusion controlled process, until the composition reaches that of the liquidus when a slower oxidat-ion process takes over. With a continuous supply of metal oxide, linear corrosion kinetics are seen [8].

A number of studies of the behaviour of silicon nitride based materials in corrosion environments have been carried out. Sodium carbonate films and sodium sulphate films and vapour [9,10] under flowing air lead to fast oxidative corrosion. Combustion gases containing high concentrations of sodium, vanadium and sulphur, and equimolar sodium sulphate-vanadate mixtures also give fast corrosion [11,12]. Pure vanadium pentoxide at

900°C is reported to be less aggressive [13]. In all cases, however, the nature of the oxidation rate controlling step, while clearly involving the corrodant and its availability, has not been identified. The basis for identifying possible protective systems is thus uncertain.

Most quantitative work on the kinetics of corrosion of silicon nitride with sodium oxide-containing materials has been with sodium carbonate and sodium sulphate. These compounds have the disadvantage of showing large initial mass losses due to the release of carbon dioxide or sulphur trioxide, and which can mask the initial oxidative mass gains used to monitor the corrosion rate. The sodium metasilicate (Na_2SiO_3) used in this programme of 'static' corrosion has no such disadvantage. It is an effective corrodant because this 1:1 Na_2O-SiO_2 composition lies well outside the liquidus envelope, and at temperatures > 795°C is a solvent for silicon dioxide. The second corrodant tested was the formally acidic vanadium pentoxide, both alone and in combination with sodium oxide as the meta-vanadate, $NaVO_3$.

2. EXPERIMENTAL

A commercial hot-pressed silicon nitride (Feldmühle, FDR) was used in the form of 7 mm cubes. This material had been densified with the aid of, primarily, yttrium oxide. A chemical analysis is shown in Table 1.

Table 1: Silicon nitride composition with respect to metallic species (expressed as their oxides).

Oxide	Y_2O_3	Fe_2O_3	Al_2O_3	CaO	MgO	ZrO_2
mass %	9.30	1.71	0.12	0.05	0.01	0.01

X-ray diffraction (XRD) and scanning electron microscopic (SEM) analyses showed that the material consisted of β-silicon nitride (predominantly), α-silicon nitride, iron silicide (Fe_5Si_3), and an yttrium-rich intergranular phase.

Following cleaning and polishing the cubes were spray coated with solutions or suspensions of corrodant using an air-brush. Some samples were corroded at temperatures in the range 900°C to 1200°C, in a simple pre-set horizontal tube furnace under flowing laboratory air (1.5 cm^3 s^{-1}) using intermittent weighing to record the oxidative mass gains. Coated samples were also suspended by platinum wire from a microbalance (CI Electronics Ltd.) in a vertical tube furnace under the same conditions. In this case mass gains due to oxidation could be recorded continuously. After oxidation surface films were examined by SEM.

3. RESULTS

Figure 4 shows mass gain data ($\Delta m/A$)^2 for materials oxidised intermittently at 900, 1000 and 1100°C in the presence of a 25 g m^{-2} sodium silicate film. Initially parabolic kinetics are closely followed, but at longer times, and especially at higher temperatures, the volatilisation of sodium oxide becomes significant and overall mass losses start to be seen. For this reason detailed analyses were made at temperatures of 1100°C, or lower.

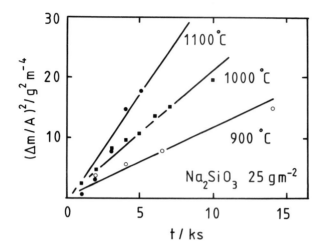

Figure 4: Mass gain data as a function of time, for silicon nitride oxidised in the presence of sodium silicate, at 900, 1000 and 1100°C.

Additions of larger amounts (to 125 g m^{-2}) of sodium silicate give faster initial corrosion rates. The initial stage parabolic rate constant is, approximately, linearly dependent on the amount of sodium silicate initially present (Figure 5).

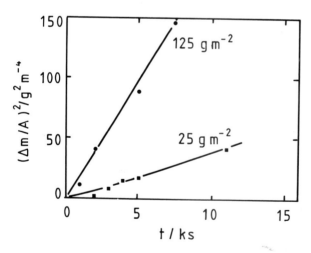

Figure 5: Mass gain data as a function of time, for silicon nitride oxidised at 1100°C in the presence of sodium silicate films.

Similar data for the oxidation of silicon nitride in the presence of vanadium pentoxide and sodium meta-vanadate (NaVO$_3$) are shown in Figure 6.

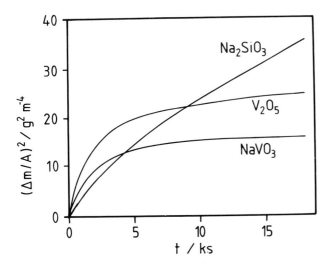

Figure 6: Mass gain data for the oxidation of silicon nitride at 900°C in the presence of sodium meta-vanadate and vanadium pentoxide.

At longer times the fast oxidative corrosion process is replaced by slower oxidation at a rate close to that for the normal oxidation reaction in air.

4. DISCUSSION

This work confirms that the 'static' corrosion of silicon nitride in the presence of a single application of corrodant occurs in two stages. The first stage is very fast, and is a consequence of the dissolution of a potentially protective silicon dioxide film (the primary oxidation product) in a liquid derived from the corrodant together with intergranular phase metal oxides leaking into it by diffusion, or by incorporation with the inwards advance of the nitride-oxide interface. The second stage corresponds to normal oxidation in the presence of a silicon dioxide film at the nitride-silicate interface. A major point of interest is that sodium oxide and vanadium pentoxide both give initially fast corrosion. This is readily explicable in terms of their ability to react with silicon dioxide to form a liquid. The model applied here assumes that oxygen diffusion through the liquid silicate film is a relatively fast process, and thus not rate-controlling, a view supported by the data shown in Figure 1. Separate studies [14] have shown that corrosion is independent of oxygen partial pressure over the pressure range 0.2 to 1 bar.

Analysis of the fast stage data in terms of the instantaneous corrosion rate shows that the rate limiting step involves a solution process which is dependent on the silicate composition. This is either the rate of solution of the primary silicon dioxide, or the solution of silicon nitride into the silicate, to be followed by its fast oxidation. Information exists about solution rates of silicon dioxide (in various forms) into sodium silicate melts of compositions similar to those found here [15]. These rates can, however, be faster than the corrosion rates seen, suggesting that the rate determining dissolution process may be that of silicon nitride itself in the silicate liquid. However, little is known about the rates of solution of silicon nitride into silicate liquids, or

about silicon nitride solubility limits. Silicates are, however, well-
studied sintering aids for silicon nitride [16], and for them to be
effective in this respect appreciable solubility in the silicate of
silicon nitride is required. Normal sintering temperatures are considerably higher (1800°C+) than those used here, but studies with fluoride
fluxes as sintering aids [17] have shown their effectiveness (and by
implication the existence of appreciable silicon nitride solubility) at
temperatures as low as 1300°C.

Corrosion data are conveniently analysed by examining the instantaneous
corrosion rate ($\dot{m}$) as a function of instantaneous silicate film composition expressed as mole fraction of SiO_2 (x_{SiO_2}). The relationship
established [14] is:

$$\ln \dot{m} = -\beta \cdot x_{SiO_2} + \gamma \qquad\qquad\qquad \ldots\ (4)$$

β and γ are constants for a given system and temperature. Figure 7 shows
these relationships for the systems examined here.

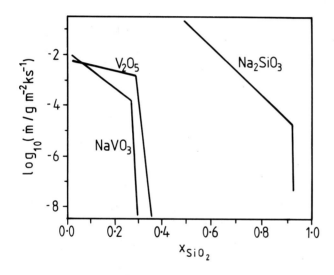

Figure 7: Corrosion rate ($\dot{m}$) as a function of mole fraction of SiO_2 in
the silicate film for corrosion at 1000°C.

Table 2 gives the β- and γ-values obtained from this treatment.

Table 2:   Equation (3) constants for silicon nitride corrosion at 1000°C.

System	Loading/ g m^{-2}	- β	$\log_{10} \gamma$ /g m^{-2} ks^{-1}
Na_2SiO_3	10.0	9.5	+ 3.4
$NaVO_3$	8.0	1.4	- 0.9
V_2O_5	4.8	5.3	- 2.2

Comparisons of the β and γ values for the three systems reported on here allows a semi-quantitative evaluation to be made of the effectiveness of these components as corroding agents. The β value is an indication of the sensitivity of corrosion rate to silicate composition; the γ value can be used as a measure of the intrinsic reactivity of the component. It is seen that, as might be expected from consideration of the phase equilibrium diagrams and the possibilities for compound formation, Na_2O is intrinsically more reactive than V_2O_5. The minimum in sensitivity corresponds to the mixed oxide system. In the long term severe corrosion damage would be expected above 695°C from vanadium pentoxide-containing atmospheres, however, because of the formation of the vanadium and silicon-containing liquid phase.

It would appear on the basis of this analysis that the scope for improvement of the corrosion resistance of silicon nitride-based materials towards corrosion may be limited. Nonetheless it is helpful to be able to express the possible susceptibility towards corrosion by a specific oxide or salt system in terms of the appropriate phase equilibrium diagrams. The most obvious step towards securing improvement in corrosion resistance is to limit the number of component oxides present in the overall system, thus raising eutectic temperatures and keeping the silicon dioxide-rich liquidus composition as distant as possible from silicon dioxide. This implies a necessity for restricting the number of oxides used as the sintering aid during the preparation of the silicon nitride, and the use of materials of high purity. In the limit this leads to consideration of the application of protective coatings, of very high purity CVD silicon nitride for example.

5. CONCLUSIONS

Both the formally basic sodium oxide and the formally acidic vanadium pentoxide accelerate the rate of oxidation of silicon nitride-based materials through their reaction with, and dissolution of, a potentially protective silicon dioxide surface film. Standard oxide phase equilibrium diagrams give useful guides to the compositions and temperatures likely to favour corrosion. The rate-controlling dissolution step is sensitive to the composition of the silicate oxidation product. This process is likely to be the rate of solution either of the primary silicon dioxide oxidation product, or of silicon nitride itself, in the liquid silicate. By using the instantaneous oxidation rate analysis, a quantitative comparison of the effectivenesses of different corrodent oxide systems is made possible.

6. ACKNOWLEDGEMENTS

The work reported here has been supported in part by the Commission of the European Communities, through JRC Petten, and through Stimulation Action Programme ST2J-0146-1-UK(D).

7. REFERENCES

1. Progress in Nitrogen Ceramics, Ed. F.L. Riley, NATO Advanced Study Institute Series E: Applied Sciences, Vol. 65, Martinus Nijhoff, The Hague (1983).

2. H. Suzuki. "The Present Attitude towards High Technology Ceramics in Japan", 255-278 in 2nd European Symposium on Engineering Ceramics, Ed. F.L. Riley, Elsevier Applied Science (London, 1989).

3. M.A. Lamkin, F.L. Riley and R.J. Fordham. "Hot-corrosion of Silicon Nitride in the Presence of Sodium and Vanadium Compounds", in Proceedings of European Colloquium: "High Temperature Corrosion of Technical Ceramics", Petten, The Netherlands, 26-28th June 1989, to be published.

4. Shi and R.A. Rapp. "The Solubility of SiO_2 in fused Na_2SO_4 at 900°C", J. Electrochem. Soc. 133 [4], 849-850 (1986).

5. N.S. Jacobson and J.C. Smialek. "Hot-corrosion of Sintered α-SiC at 1000°C", J. Amer. Ceram. Soc. 88 [8], 432-39 (1985).

6. Phase Diagrams for Ceramists, American Ceramic Society, Columbus, OH (1964).

7. G. Gauthier, F.L. Riley and R.J. Fordham. "Sodium Salt Corrosion of an Yttrium Oxide Densified Silicon Nitride", to be published.

8. M.I. Mayer and F.L. Riley. "Sodium Ion Assisted Oxidation of Reaction-Bonded Silicon Nitride", Proc. Brit. Ceram. Soc. 26, 251-64 (1978).

9. M.I. Mayer and F.L. Riley. "Sodium Assisted Oxidation of Reaction-Bonded Silicon Nitride", J. Mater. Sci. 13, 1319-1328 (1978).

10. G. Gauthier, M.A. Lamkin, F.L. Riley and R.J. Fordham. "The High Temperature Corrosion of Silicon Nitride", The Institute of Ceramics Proceedings 39, 55-59 (1987).

11. S. Brooks and D.B. Meadowcroft. "The Corrosion of Silicon Based Ceramics in a Residual Oil Fired Environment", Proc. Brit. Ceram. Soc. 26 237-50 (1978).

12. S. Brooks, J.M. Ferguson, D.B. Meadowcroft and C.G. Stevens. "Corrosion Above 700°C in Oil-Fired Combustion Gases", pp. 121-138 in Materials and Coatings to Resist High Temperature Corrosion, Eds. D.R. Holmes and A. Rahmel, Applied Science Publishers Ltd. (London, 1978).

13. S.C. Singhal. "Corrosion-Resistant Structural Ceramic Materials for Gas Turbines", pp. 311-334 in Proc. Gas Turbine Materials in the Marine Environment, Eds. J.W. Fairbanks and I. Macklin, Columbus, Ohio, USA (1975).

14. G. Gauthier, R.J. Fordham and F.L. Riley. "The Corrosion of Silicon Nitride by Sodium Silicate", to be published.

15. M. Cable and D. Martlew. "The Effective Binary Diffusivity of Silica in Sodium Silicate Melts: A Review and Recommendation", Glass Technology 25 [6], 270-276 (1984).

16. V. Vanedenede, A. Leriche, F. Cambier, H.Pickup and R.J. Brook. "Sinterability of Silicon Nitride Powders and Characterization of Sintered Material". In Non-oxide Technical and Engineering Ceramics, Ed. S. Hampshire, Elsevier (London, 1986), pp. 53-67.

17. J.R. Oswald, F.L. Riley and R.J. Brook. "Accelerated Densification of Silicon Nitride Using a Fluoride Flux", British Transactions and Journal 86, 81-84 (1987).

Thermal shock evaluation of high temperature structural ceramics

Yasunobu Mizutani

Technical Research Institute , TOHO Gas Co., Ltd.
507-2 Shinpo-machi Tokai-city Aichi-pref. 476 JAPAN

ABSTRACT : A new experimental method is proposed for determining
thermal shock resistance parameter of ceramics under thermal and
mechanical combined stress conditions. A three-point bending specimen is
preliminary loaded and heated on its one surface of constant heat flux,
and the time to fracture is measured. The experimental conditions are
theoretically calculated by analytical solution of heat conductive
equation and finite element method. The experimental data are useful for
designing for the application of ceramics.

1. INTRODUCTION

It is water quenching method that has been conventionally used by
material manufacturers for their thermal shock tests. This method, however,
needs a number of specimens, and further, the thermal shock resistance
value is obtained through trial and error. On the other hand, at the user's
side, thermal shock resistance is evaluated on the actual application
conditions on a single specimen (such as parts of appliances). In this case,
whenever the application conditions are changed, even if the change is
minimal, the evaluation has to be renewed. Further, because the evaluation
is made on one parts, the other parts are left not evaluated. (And further,
in this case, dispersion of the material is not considered at all.)

In this paper, the author presents a new and simple thermal shock
evaluation taking these problems into good considerations. For studying
the thermal fracture of ceramics under thermal stress and mechanical stress
combined, a three-point bending bar-type specimen was used in the thermal
shock test and observed some information as described hereinafter.

2. CONVENTIONAL THERMAL SHOCK EVALUATION

Generally, water quenching (a strength measuring method after
quenching into the water) has been used for thermal shock tests on
high temperature structural ceramics. The water quenching, however, has
some problems as follows :

1)   Because a large number of specimens are required, the difference in
measurement conditions is likely to cause errors, and further the test has
to be cycled through trial and error.

2)   Because of heat transfer by means of boiling water (gas/liquid

mixing), heat transfer coefficient at the boundary layer with the specimen can not be constant (and therefore theoretical estimation is difficult).

3)   Because the Biot's number (severity of thermal shock) varies according to the size and shape of the specimen, data also varies according to the test conditions.

4)   Because the thermal shock is not tested under the actual application conditions (often with severer thermal shock condition), conclusion can not be made simply in selecting the material.

To avoid these problems, recently there has been a case where laser beam heating thermal shock test is used instead. This method, however, inconveniently requires larger test equipment.

3. OUTLINE OF NEW THERMAL SHOCK TEST

One of the important problem with the water quenching is that thermal shock conditions can not be controlled. This problem can not be solved, even if the quenching media is changed. For this reason, the thermal shock was tested by heating the resistor to recreate the circumstances in which thermal shock may be observed and the conditions may be controlled.

In the thermal shock test which the author describes in this report, a conductive paste heating element is used for heating the specimen, and thermal stress is caused to the specimen by applying current to the heating element. This method allows only one side of the specimen to be heated rapidly and evenly by applying a constant heat flux.

Two type of specimens are considered. One is shaped into a ring and the other into a bar as shown in Figure 1. The ring-type specimen allows the evaluation of thermal fracture caused by thermal stress alone, while the bar-type specimen allows the evaluation of thermal fracture caused by both thermal stress and mechanical stress combined.

These thermal shock test is featured as follows :

1)   Thermal shock   conditions can be controlled   easily because of electrical heating.

2)   Thermal   shock   resistance parameter can be obtained using a few   specimen.

3) Constant heat flux conditions, and theoretical analysis can be easily made.

4)   Bar-type specimen restricted at   three   points   allows   the evaluation   of   combined   stress made   of   thermal   stress   and mechanical   stress   generated   by applying   static   load   to   loading points.

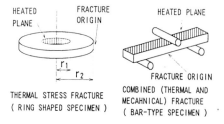

HEATED PLANE   FRACTURE ORIGIN

HEATED PLANE

r_1

r_2

THERMAL STRESS FRACTURE
(RING SHAPED SPECIMEN)

FRACTURE ORIGIN

COMBINED (THERMAL AND MECAHNICAL) FRACTURE
(BAR-TYPE SPECIMEN)

Fig.1   Schematics of test specimens

4. THEORETICAL CONSIDERATIONS

4.1 RING-TYPE SPECIMEN

When the inside wall of a thin ring-type specimen with inside radius r_1 and outside radius r_2 is heated by a constant heat flux q and the outside wall is insulated, thermal stress can be obtained by the following equations:

Equation of thermal stress :

$$\sigma_\theta^* = (1/\xi^2)[k_2 \int_a^1 T \xi \, d\xi + \int_a^\xi T \xi \, d\xi] - T \tag{1}$$

$$\sigma_\theta^* = \sigma_\theta / E\alpha(\theta f - \theta i), \; \kappa_2 = (\xi^2 + a^2)/(1 - a^2)$$

Equation of heat conduction :

$$\partial^2 T / \partial \xi^2 + (1/\xi) \partial T / \partial \xi = \partial T / \partial \xi \tag{2}$$

Initial conditions : $T = 0$ at $\eta = 0$

Boundary conditions : $\partial T / \partial \xi = -Q$ at $\xi = a$

$$\partial T / \partial \xi = 0 \quad \text{at} \quad \xi = 1 \tag{3}$$

where
$$T = (\theta - \theta i)/(\theta f - \theta i), \eta = \kappa t / r_2^2, \; \xi = r / r_2$$
$$Q = q r_2 / \lambda (\theta f - \theta i), \; a = r_1 / r_2$$

Q is constant independently of the temperature. When the equation of thermal stress is obtained from the above answer and $\xi = 1$ is substituted for the answer, the following result can be obtained :

$$\sigma_\theta^*(1, \eta) = (Q/2k_1)[\{2/(1-a^2)\}a^2]na$$

$$+(1+a^2)/2] + 2Q \sum_{n=1}^{\infty} (Cn/Bn) \exp(-\delta n^2 \eta) \tag{4}$$

$An = \{J_1(\delta n) Y_0(\delta n \xi)$

$\quad -J_0(\delta n \xi) Y_1(\delta n)\} / \delta n^2$

$Bn = \{J_0(\delta n) - J_1(\delta n)/\delta n\} Y_1(\delta na)$

$\quad + aJ_1(\delta n)\{Y_0(\delta na) - Y_1(\delta na)/\delta na\}$

$\quad - a\{J_0(\delta na) - J_1(\delta na)/\delta na\} Y_1(\delta n)$

$\quad - J_1(\delta na)\{Y_0(\delta n) - Y_1(\delta n)/\delta n\}$

$Cn = \{a(1+a^2)/(1-a^2) \cdot [J_1(\delta na) Y_1(\delta n)$

$\quad -Y_1(\delta na) J_1(\delta n)] - a[J_1(\delta n) Y_1(\delta na)$

$\quad -Y_1(\delta n) J_1(\delta na)] - \delta n[J_1(\delta n) Y_0(\delta n)$

$\quad -J_0(\delta n) Y_1(\delta n)]\} / \delta n^3$

$k_1 = 1/a - a$

where δn is the "n"th positive root which can satisfy the following equation : $J_1(\delta a)/J_1(\delta) = Y_1(\delta a)/Y_1(\delta)$

4.2 BAR-TYPE SPECIMEN (INFINITE PLATE)

When an infinite plate is heated or cooled from one side, the equation of thermal stress and equation of heat conduction are as follows :
Equation of thermal stress :

$$\sigma^* = -T + \int_0^1 T \, d\xi + (12\xi - 6) \int_0^1 (T\xi - \frac{1}{2}) \, d\xi \tag{5}$$

Equation of heat conduction :

$$\partial^2 T / \partial \xi^2 = \partial T / \partial \xi \tag{6}$$

Initial conditions : $T = 0$ at $\eta = 0$

Boundary conditions : $\partial T / \partial \xi = 0$ at $\xi = 0$

$- \partial T / \partial \xi = - Q$ at $\xi = 1$ (7)

Where $\xi = x / \ell$, $\eta = \kappa t / \ell^2$, $T = (\theta - \theta_i) / (\theta_f - \theta_i)$, $Q = q \cdot \ell / \lambda (\theta_f - \theta_i)$
When the equation is solved, the equation of temperature distribution is as
follows :

$$T (\xi , \eta) = Q \{ \eta + \frac{1}{2} \xi^2 - \frac{1}{6} - 2 \sum_{n=1}^{\infty} \frac{(-1)^n \cdot \exp (- \delta n^2 \eta) \cdot \cos \delta n \xi}{\delta n^2} \}$$ (8)

When nondimensional thermal stress is $\sigma^* = \sigma / E \alpha (\theta_T - \theta_i)$ can be
expressed by the following equation :

$$\sigma^* = Q \{ - \frac{1}{2} \xi^2 + \frac{1}{6} + 2 \sum_{n=1}^{\infty} \frac{(-1)^n \cdot \exp (- \delta n^2 \eta) \cdot \cos \delta n \xi}{\delta n^2} \}$$ (9)

As fracture origin the under side of the specimen , when $\xi = 0$ in
Equation (9) , the following equation can be obtained :

$$\sigma^*_{\xi=0} = Q \{ \frac{1}{6} + 2 \sum_{n=1}^{\infty} \frac{(-1)^n \cdot \exp (- \delta n^2 \eta)}{\delta n^2} \}$$ (10)

The result of the above
calculation (Equation (8) to
(10)) can be shown as Figure 2
to 4 respectively. Thermal
stress increases as the time
passes and then indicate a
constant value.

4.3   CALCULATION OF THERMAL
SHOCK RESISTANCE PARAMETER

Fourier's number η is
obtained from the thermal
diffusivity of the specimen κ
and the time until the
specimen fractures t , and
σ^* / Q is obtained from
Equation (10). Because both
σ^* and Q contain $\Delta \theta$ in
their denominator, if $\Delta \theta$ is
eliminated from σ^* and Q ,
referred to as σ^{**} and Q'
respectively, their ratio will
remain the same .   That is,
if σ^* / Q is multiplied by Q'
obtained from the calorie
supplied in the test, thermal
shock resistance parameter
σ^{**} ($= R_t = S / E \alpha$) will
be obtained. In an actual
experiment, not all the

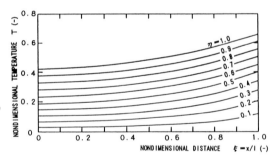

Fig.2   Temperature distribution of the
plate changes in Fourier's numbers

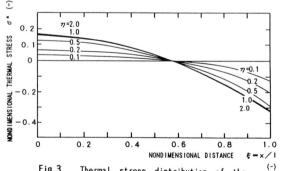

Fig.3   Thermal stress distribution of the
plate in various Fourier's numbers

calorie supplied by the heating element in the test is transferred to the specimen, Q should be corrected by the measured temperature of the outside wall of the specimen and obtain Q_{EXP}. In the case of combined stress where preliminary load is applied by three-point bending, when the mechanical stress is σ, $R_L = \sigma / E\alpha$ is obtained for facilitating the calculation, and the thermal shock resistance parameter is obtained from $R = R_t + R_L$.

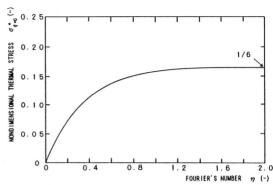

Fig.4 Thermal stress changes in Fourier's number

4.4 ANALYSIS OF THERMAL STRESS BY FINITE ELEMENT METHOD

Then, thermal stress is analyzed by finite element method on both the conditions of this test. The finite element method is used for obtaining the solution by dividing the analysis object into finite element, and solving each simultaneous equations.

The finite element method program, used for this test can analyze the two-dimensional unstable thermal stress, consists of the followings :

(1) Element type : Triangular primary element and eight-joint isoparametric element

(2) Analysis details : Thermal conduction analysis, thermal elasticity deformation analysis, thermal stress analysis.

(3) Input data : Mesh data, thermal boundary conditions, elastic boundary conditions.

(4) Output mode : Temperature distribution map, thermal history, deformation map, normal stress map, stress distribution map.

(5) Others : Process selection on menu screen.

Figure 5 shows the calculation

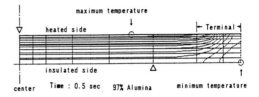

Fig.5 Temperature distribution of the specimen caluculated by FEM

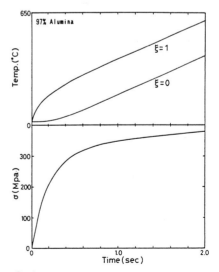

Fig.6 Changes of the temperature and thermal stress with time calculated by FEM

result of the temperature distribution in the case of the bar-type specimen.

Because the terminal part is not heated, a partial fall in temperature is witnessed. But the effect of this temperature fall on the fractured surface is small, and also the effect of the thermal shock resistance parameter on the measured value seems to be small.

Figure 6 shows the progress in the temperature and thermal stress. These curves follows the theoretical analysis (obtained by the heat conductive equations) , providing that each condition of this test follows the stress equation which is theoretically obtained.

5. THERMAL SHOCK TEST ON BAR-TYPE SPECIMEN UNDER COMBINED STRESS

5.1 EXPERIMENTAL PROCEDURE

Bar-type specimens used for JIS (Japan Industrial Standards) R-1601 "Bending strength test" were used. Table 1 shows the thermal and mechanical properties of the specimen used for the test.

Table 1 **Mechanical and thermal properties of the tested materials**

	Size b×l×L (mm)	Bending strength (MPa)	Young's modulus (GPa)	Thermal expansion coef. ($\times 10^{-6}$/K)	Thermal conduc -tivity (W/mK)	Specific heat (J/KgK)
Steatite	3×4×51	135	70	6.2	2.2	925
97% Alumina	3×4×48	320	180	6.7	10	800
TM-D (99% Alumina)	3×4×42	480	250	6.8	15	800

To make the surface roughness uniform, the specimen surface to be fractured by thermal shock was polished by 3 μ m diamond slurry, and measured in size and dry weight. Ruthenium paste (1 Ω / $\square$) , a heating element was screen printed to one side of the specimen. and then dried and fired at 850 °C . After measuring the thickness and dry weight of the resistor, conductive Ag paste was applied, dried, and then fired in the same way. Then, the electrode lead wire was fixed by high-temperature soldering and the resistance value was measured. The measured value was about 5 to 15 Ω . The prepared specimen was fixed by the three-point bending jig. After the specified static load (preliminary load) was applied to the specimen, and one side of the specimen was heated quickly to cause thermal shock and fracture to the specimen.

Because the fracture was occurred within as short as several seconds, data for the current, voltage, and time to fracture were entered in the personal computer through the AD converter for measurement. To monitor the reaction caused by the static load stress and thermal stress, the load at the three-point bending upper restriction point was measured using the load cell.

Figure 7 shows the components of the equipment used the test.

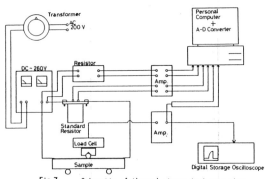

Fig.7   Schematic of thermal stress test apparatus

6. RESULTS AND DISCUSSION

6.1 EFFECT OF HEAT FLUX

In order to understand the relation between the calorie supplied in the test and the measured thermal shock resistance parameter, heat flux is changed variously and measured. The result of Steatite is shown in Figure 8.

The measured thermal shock resistance parameters showed constant values over a wide range, excepting the range where Q is small. This proves that precise measurement can be made independent of the testing conditions. The reason why the measured values are small within the range where Q is small is that the time required for causing the fracture is long and strength is lowered due to the SCG (slow crack growth) effect.

In other words, to obtain thermal shock resistance parameter in this test, voltage to be applied should be adjusted to control thermal stress. That is by increasing Q' considerably and reducing the time required for causing fracture to the specimen (Fourier's number), the effect of SCG can be avoided and precise thermal shock resistance parameter can be measured.

6.2 EFFECT OF MECHANICAL STRESS

Figure 9 shows the result of the test conducted by varying the ratio of thermal stress to mechanical stress on 97% alumina and TM-D. The X-axis shows the ratio of thermal shock coefficient R_L to mechanical stress coefficient R_t . In both cases, R was almost constant value over the mechanical stress range. Therefore, it is provided that thermal shock resistance parameter can be measured under combined stress as well.

6.3 DISTRIBUTION OF MEASURED DATA

Assuming that thermal shock fracture is attributed to cracks existed inside the material, the obtained thermal shock resistance parameter and the fracture strength data obtained through three-point bending test are Weibull plotted respectively as shown Figure 10 .

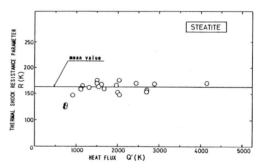

Fig.8   The relation between thermal shock
resistance parameter R and heat flux Q'

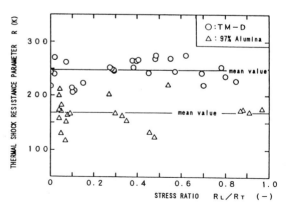

Fig.9   The relation between thermal shock
resistance parameter R and stress rate R_L/R_T

Each data is obtained follows the Weibull distribution, and the Weibull modulus m obtained from thermal shock test is slightly lower than the value obtained from the material strength data. This result can be explained from the difference between the geometric factor of each cases. From these results, reliable thermal shock resistance parameter may be measured using several specimens.

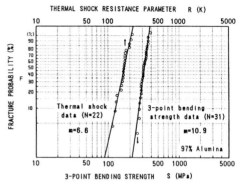

Fig. 10   Weibull plots of the thermal shock resisitance parameter and 3-point bending strength

6.4   STRESS DISTRIBUTION OF THE SPECIMEN

Generally, thermal stress and mechanical stress are often different from each other in stress distribution inside the material as this test shows unexceptionally. Figure 11 shows stress distribution in the direction of thickness inside the plate in three cases - (1) thermal stress only (heated one side), (2) combined stress, (3) mechanical stress (three-point bending). This figure shows the fact that the stress distribution and geometric factor are differ among each cases.

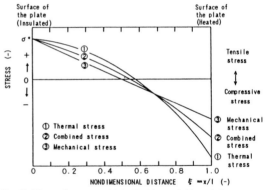

Fig. 11 Thermal stress distribution of the plate in various stress conditions

Further, the fractured specimen conditions are compared on both the cases. In the case of (1), the specimen is fractured so severely that a large number of fragments are caused.(Figure 12)

On the other hand, in the case of (2), fracture is less severe than that of (1), and in most cases, the specimen is fractured into two pieces. This can be explained from the fact that elastic strain energy is larger than thermal distribution shown in Figure 11 (the area with oblique lines is larger).

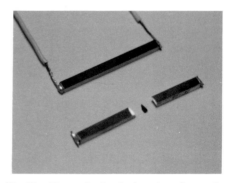

Fig. 12   Thermal shock fractured specimen

7. CONCLUSIONS

Under the thermal shock conditions that resistor paste is used as a heating element and one side of the specimen is rapidly heated, the thermal shock resistance parameter of bar-type specimen was measured making use of the equation of thermal stress to obtain the analytic solution, and the following conclusions were obtained :

(1) The measured thermal shock resistance parameter is not dependent on the size of thermal flux over a wide range and is scarcely affected by the test conditions.

(2) Reliable thermal shock resistance parameter can be obtained in this test on several specimens from the Weibull statistics of the test data. This is very significant because water quenching needs a large number of specimens.

(3) Thermal shock resistance can be evaluated in this test under thermal stress and mechanical stress combined.

(4) Thermal shock resistance parameter obtained in this test can be reproduced and therefore can be used sufficiently for the design value of thermal stress analysis.

*** ACKNOWLEDGEMENTS ***

The author owes the accomplishment of the test to the guidance extended by Prof. Manabu Takatsu and Mr.Kamiya, Assistant to Prof. Takatsu, Inorganic Material Course of Material Engineering Department, Nagoya Institute of Technology, and the cooperation extended by students under Prof. Takatsu (Mr.Kamiya is now working for Nagoya University as an assistant). The author would like to express his sincere gratitude to them.

*** NOMENCLATURE ***

α : thermal expansion coefficient $[K^{-1}]$
λ : thermal conductivity $[W/m\ K]$
σ : thermal stress $[Pa]$
σ^{*} : nondimensional thermal stress $[-]$
θ : temperature $[K]$
   (subscript i or f indicate initial or final condition, respectively)
T : nondimensional temperature $[-]$
η : Fourier's number $[-]$
t : time $[sec]$
κ : thermal diffusivity $[m^2/sec]$
ξ : nondimensional distance $(=x/1)$ $[-]$
x : distance in thickness direction $[m]$
1 : thickness of the specimen $[m]$
L : length of the specimen $[m]$
r : radius of ring specimen $[m]$
q : heat consumption $[W/m^2]$
Q : nondimensional heat flux $[-]$

Q' : parameter of heat flux [K]

E   : elasticity [Pa]

S   : strength [Pa]

R   : thermal shock resistance parameter [K]

J_n : Bessel function of first kind
(0 or 1 indicate zero or first order, respectively)

Y_n : Bessel function of second kind
(0 or 1 indicate zero or first order, respectively)

Paper presented at Conf. on Ceramics in Energy Applications, Sheffield, April 1990
Session 2A

Ceramic fibre plates as flame holders

[1]Flemming Andersen, [1]Michael Andersen, [2]Jan Jensen, [1]Sven Hadvig.

[1]Laboratory of Heating and Air Conditioning, Technical University of Denmark, Building 402, DK-2800 Lyngby, Denmark.
[2]Danish Gas Technology Centre a/s, Dr. Neergaardsvej 5A, DK-2970 Hoersholm, Denmark.

ABSTRACT: In this paper some measured results from fibre burners made of inexpensive ceramic material originally intended for high temperature insulation are presented. A hexagonal cylinder was made of flat fibre material already availabel on the market and used as a burner in an ordinary circular cylindrical heater. The main features of this fibre burner are low emission of nitrogen oxydes and high radiation from the surface. The fibre burner used in a CO_2-generator is also described. Further the cause of the low nitrogen oxide production is described.

1. INTRODUCTION

The results in this paper are all a part of a research program concerning fibre burners. The program, which was initiated in 1987, is funded by the Danish Ministry of Energy and Danish Gas Technology Centre.

At the moment there is an increasing interest for producing energy with the least possible pollution. One of the methods for reducing the emission of nitrogen oxide in areas where natural gas is availabel is to use fibre burners. The high radiative heat transfer causes a lower flue gas temperature and thus a lower nitrogen monoxyde emission since the formation is extremely dependent of the flue gas temperature.

It is expected that also the heater can be produced at a lower cost price since more energy is transferred per square meter heater surface due to the higher radiation. Therefore a smaller heater surface is needed for a given energy output.

2. COMBUSTION PRINCIPLE IN FIBRE BURNER

In Figure 1 is shown the principle for combusting natural gas by using a fibre burner. A homogeneous mixture of natural gas and air enters the fibre layer at the left (at $X=0$). The mixture is heated by the fibres which before the combustion zone are warmer than the gas. Heat is transferred to the left by heat conduction in the fibres, radiation between the fibres and by conduction in the gas. When the gas is heated to a high temperature, the chemical combustion reaction starts (at $X=L_1$) and heat is produced in the exothermal processes. The combustion zone ends (at $X=L_2$). In a large part of the combustion zone and to the right of the zone the gas has a higher temperature than the fibres and heat is transferred to the fibres by Newton's cooling law.

The fibres are heated to such a high temperature that the fibre burner surface is glowing red. The surface temperature is approx. 950°C-1050°C, depending on the chemical composition of the fibres, the excess air, the specific power input, the chemical composition of the natural gas and the porosity of the fibre layer.

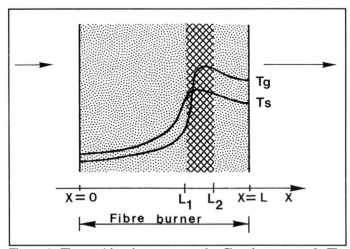

Figure 1. The gas/air mixture enters the fibre layer at x=0. The combustion starts at $X=L_1$ and ends at $X=L_2$. The hot flue gas leaves the fibre layer at $X=L$. T_g and T_s are the temperatures of the gas and the fibres, respectively.

Since the fibres are in good thermal contact with the gas, the temperature of the flue gas inside the burner when leaving the burner is lower than by many other types of burners. The production of the NO is just like many chemical reactions nearly exponential dependant of the temperature, so when the temperature is decreased, the production of the NO_x is reduced also. This is the main reason why fibre burners produce less NO_x than conventional burners.

3. DESCRIPTION OF THE HEXAGONAL FIBRE BURNER

The material for the burner was chosen after experiments made on a small test rig. This test rig and some of the results are described in Jensen et al. (1989) and in more detail in H. Soerensen and Fl. Andersen (1989). On this rig was measured emission of CO, C_xH_y and NO_x. Further the pressure drop over the fibre plates and the radiation from the hot burner surface were measured.

The fibre plates were not only tested with natural gas from the North Sea but also with a gas containing 35% CH_4 and 65% H_2 (G22 in DIN 3362). The last mentioned gas has a higher flame velocity and a higher net heating value. By using this gas the fibres are effected thermally more severely and thus a sort of accelerated test was made.

From the above measurements the fibre material shown i Tabel 1 was chosen to the 600 kW burner. The burner was constructed from perforated plates of steel as it is seen from Figure 2. On this steel frame the plane fibre plates with a porosity of approx. 90% was glued. Figure 2 shows a section in the hexagonal burner. Figure 3 shows a view of the hexagonal burner. The results concerning the hexagonal burner, are taken from Michael Andersen (1989), where a further description of the burner is given.

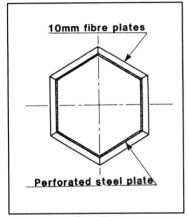

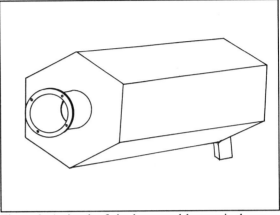

Figure 2. The figure shows a sketch of a section in the hexagonal burner.

Figure 3. A sketch of the hexagonal burner is shown.

Tabel 2 shows the chemical components in the fibres used for the burner. The fibres are made by a Danish producer from mainly diabas stones, and the batts made of these fibres are normally used as a thermal insulation material, fire protection insulation and for noise reducing purposes.

Burner area	1.88 m²
Nominel Burner effect	600 kW
Porosity	90%
Average fibre diameter	3.5 μm
Thicknes of fibre plates	10 mm
Density of fibre	2860 kg/m^3

Tabel 1. Data of the burner and the used fibre material.

SiO_2	47.5 %
Al_2O_3	13.0 %
TiO_2	1.5 %
Fe_2O_3	0.5 %
FeO	7.0 %
CaO	16.0 %
MgO	10.5 %
MnO	0.5 %
Na_2O	2.5 %
K_2O	1.0 %

Tabel 2. The chemical components in the fibres.

4. MEASUREMENTS AND RESULTS

The total radiation to the boiler wall was measured with hemispherical radiation meters (hollow ellipsoide pyrometers). The temperature and the water flow inside the heater were measured by Pt-100 thermometers and magnetic flow meters, respectively. The NO_x, CO, C_xH_y and the O_2 are all measured on dried flue gas. NO_x is measured by a chemiluminiscent analyser, CO by a infrared analyser, C_xH_y by a flame ionization detector and O_2 by a paramagnetic analyser. Figure 4 and Figure 5 show the main points of the measurements. The symbols in these figures, i.e. T, P and FL, mean thermometer, manometer and flowmeter, respectively.

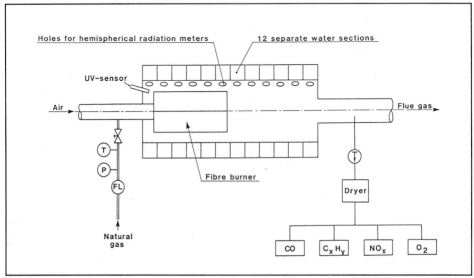

Figure 4. The drawing shows the hexagonal cylindrical fiber burner inside a circular heater. The main points of measurements are shown. The water sections are numbered from the left, from 1 to 12.

The measured data were collected and stored by a HP-computer with a datalogger. All the data shown in the following figures were measured under stationary conditions in the heater/burner-system.

The emission of the mentioned gases and the hemispherical radiation were measured as functions of the input power and the excess air ratio.

The distribution of the power absorbed by the heater was calculated from the measured water temperatures and flows.

These results are shown in Figure 6 - Figure 10.

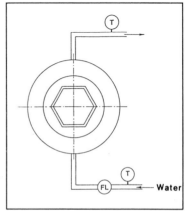

Figure 5. A section in the burner and heater is shown.

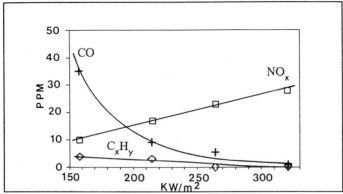

Figure 6. The emission of NO, CO and C_xH_y in the flue gas leaving the heater is shown as a function of the specific power input by 10% excess air.

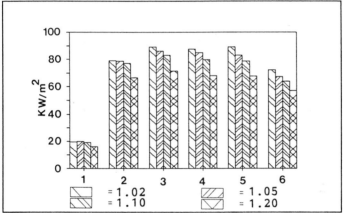

Figure 7. The hemispherical radiation at the heater wall is shown as a function of the section number and the excess air.

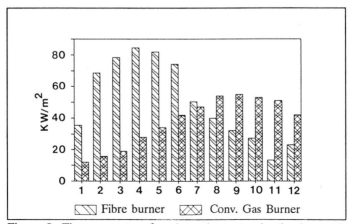

Figure 8. The power transferred to the water in the heater is shown as a function of the section number.

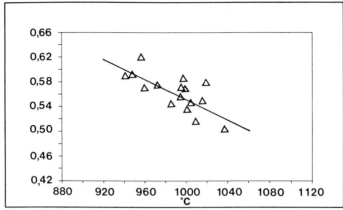

Figure 9. The emissivity is shown as a function of the surface temperature.

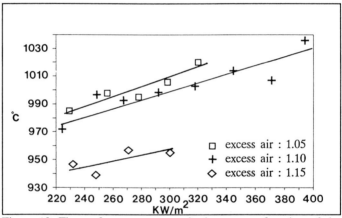

Figure 10. The surface temperature is shown as a function of the excess air and the specific power input.

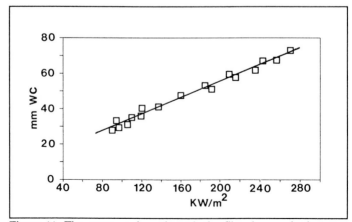

Figure 11. The pressure drop through the fibre burner is shown as a function of specific input power by 10% excess air.

5. PRACTICAL APPLICATIONS

It is a well known fact that by raising the CO_2-content in the atmosphere of greenhouses, a higher production of plants, flowers and vegetabels is optained. However, if the NO_x is too high this may have a negative influence on a least some of the plants causing slower growth or poorer quality of the plants. At present the CO_2 is supplied either by burning different kinds of fuels inside the green houses or by supplying CO_2 from pressurized bottles or tanks.

The fibre burner is regarded as suitabel for producing CO_2 for greenhouses because of the low content of NO_x and CO in the flue gas. The Laboratory has constructed such a CO_2 generator, which has been succesfully tested in a greenhouse for more than 6 months. A sketch of the generator is shown in Figure 12. The combustion air from the compressor (1) is mixed with natural gas (2). The mixture enters the fibre layer (3) and burns in the surface of the fibre layer. The flue gas is mixed with the green-house atmosphere by the fan (4). The gas and air valves are kept open, when the UV-sensor (5) detects combustion.

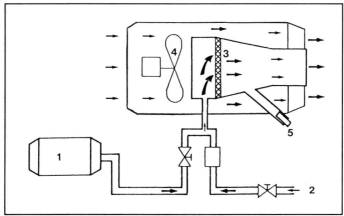

Figure 12. The sketch shows the CO_2-generator for greenhouses.

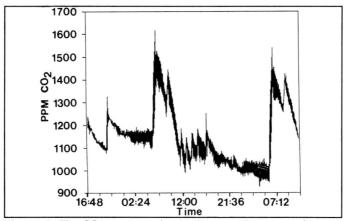

Figure 13. The CO_2 concentration in the green house as a function of the time of the day

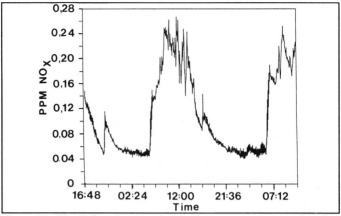

Figure 14. The NO_x concentration in the greenhouse as a function of the time of the day.

The average NO_x-concentration is approx. 0.1 ppm and the CO_2 is approx. 1200 ppm in the greenhouse atmosphere. The concentrations of CO_2 and NO_x as a function of the time in the day are shown in Figure 13 and Figure 14 respectively and were measured by Danish Technology Institute.

A Danish producer has taken interest in CO_2-generators and has constructed and produced a small series of the generators for further testing in greenhouses. The company expects to have a final generator for commercial sale in 1990.

The burners can be used both in large heaters for district heating and in domestic heaters. The main advantages of this burner are low NO_x-emission and high radiation. At the moment the main disadvantage is a high pressure drop. This is a disadvantage because the expenses for the energy to the compressor is almost directly proportional to the pressure drop through the fibre burner. Also the price of the compressor is dependent on the pressure, since compressors or fans for low pressure are very inexpensive and multi stage or piston compressors are more expensive.

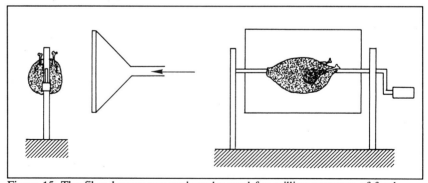

Figure 15. The fibre burner can perhaps be used for grilling purposes of food.

6. DISCUSSION

From the emission graphs it is seen that the NO_x-emission is very low compared to conventional gas burners. The CO and the C_xH_y is surprisingly high compared to the measurements on the small test rig described in J. Jensen et al. (1989). These high values are probably caused by the fitted corners of the burner, where poor combustion takes place.

The total hemispherical radiation measured along the heater is quite uniform as it is seen from Figure 7. The power transferred to the heaters cooling water is also quite high and higher than the heat transfer measured at the same heater with a conventional gas burner, see Figure 8 and S. Hadvig et al. (1988).

From Figure 7 is seen that the hemispherical radiation decrease when the excess air increase. This effect is due to the fact that the higher mass flow has to be heated by the same net heating value. The flue gas optains a lower temperature and hence also the fibre surface temperature is lower.

Figure 11. shows the pressure drop through the hexagonal fibre burner. The pressure drop is quite high. It is seen to be nearly linear dependant of the input power as it is expected from Darcy's law.

In H. Soerensen and Fl. Andersen (1989) is described a method to calculate the emissivity of the fibre burners from measured hemispherical radiation, gas temperature, wall temperature and fibre burner surface temperature. The last mentioned temperature was measured by using a two-colour pyrometer. The measurements were all done at the Laboratory's flat research heater. The graphs are shown in Figure 8 and Figure 9. The burner surface temperature and emissivity are the most important parametres, when heat transfer calculation in the combustion chamber are made.

7. RESEARCH AT THE LABORATORY

At the moment a closer cooperation with a fibre company is establised with the purpose of producing circular cylindrical burners without corners, at which problems can occur.

Further research and development performed to decrease the pressure loss over the burner. It is important to minimize this presure drop to lower the power expenses to the compressor feeding the combustion air to the burner.

Also mathematical modelling of the heat transport inside the fibre layer is at the moment a subject for research. An introduction to this area is given in Fl. Andersen (1989).

8. CONCLUSION

Although the material was testet carefully on a small rig before the 600 kW burner was made there remained problems with the corners of the burner, where the fibre plates were fitted together. However, it is our strong beleif that it is possible to create such a large burner in a very inexpensive way by using temperature resistant insulation material. Special attention must be given the corners of the burner.

From the emission graphs is seen that the NO_x in the flue gas is approx. 23 ppm which is considerably lower in flue gas from ordinary gas burners.

It is clear that pollution of nitrogene oxide from natural gas fired domestic heaters and from heaters used in district heaters can be drastically diminished by using fibre burners.

9. ACKNOWLEDGEMENT

We would very much like to thank the Danish Ministry of Energy for funding and appreciating this project.

10. REFERENCES

Andersen, Flemming, Analysis of natural gas fired ceramic fibre Burner, Part II, Theoretical study, Technical University of Denmark, Laboratory of Heating and Air Conditioning. Paper presented at XI-th Internaltional Symposium on Combustion Processes, Miedzyzdroje, Poland, 1989.

Andersen, Michael, Fiberbraender, M.Sc. thesis, Technical University of Denmark, Laboratory of Heating and Air Conditioning, 1989.

Hadvig S, Madsen O, Jensen J. Enviromentel and heat transfer features of a fibre burner and conventional burners for gas and oil respectively, Technical University of Denmark, Laboratory of Heating and Air Conditioning. Paper presented at the 17th World Gas Conference, Washington D.C., June 1988.

Jensen J, Krighaar M, Andersen Fl, Hadvig S, Analysis of natural gas fired ceramic fibre burner, Part I, Experimental Study and Practical Applications, Danish Gas Technology Centre. Paper presented at XI-th Internaltional Symposium on Combustion Processes, Miedzyzdroje, Poland, 1989.

Soerensen H, Andersen Fl, Integreret Fiberbraender og Kedelsystem, M.Sc thesis, Technical University of Denmark, Laboratory of Heating And Air Conditioning, 1989.

Use of ceramics in a low NO$_x$ single ended recuperative radiant tube

Laurent SCRIVE, Jean-Pierre CASSAGNE

Research and Development Division, GAZ DE FRANCE

ABSTRACT : The appearance on the market of high temperature ceramics at reasonable cost has provided an excellent opportunity for equipment designers. In the past, one of the major obstacles to energy recovery was the high temperatures to which certain burner, radiant tube or exchanger parts were subjected. Now, with the use of heat resistant ceramic parts, efficient, reliable equipment can be made at a reasonable price, provided that metal alloys and ceramics are combined in such a way that the physical characteristics of each are made compatible.

1. INTRODUCTION

The gas-fired high temperature radiant tube is used in a wide range of indirect heating furnaces.

The most efficient tubes are the single ended radiant tubes. They include staged combustion tubes, housed inside an outer tube, and radiant recirculation tubes in which a jet burner draws in a fraction of combustion products at the burner nozzle. The high speed recirculation of flue gases between the inner tube and the annular space between the inner and outer tubes results in a more uniform temperature along the two tubes.

As it is more difficult to obtain correct staged combustion over time in a conventional radiant tube, especially if the calorific value of the gas varies by several percent, we therefore opted for principle of the single-ended recirculation radiant tube.

The development of this novel heating equipment was made possible by the appearance on the market of new materials and equipment at reasonable prices. They include :

- industrial ceramics, capable of withstanding much higher operating temperatures than refractory steels,

- self-adapting control systems for regulation of furnace zones by pulse modulation,

- gas and air solenoid valves operating reliably in on/off mode (90 % are able to withstand 10 million cycles, according to the manufacturer, after several series of statistical tests).

These new technologies, associated with accurate identification of the functions required by potential users using the value analysis method, and assessment of the effects of combustion on the environment, have enable us to design and develop the high performace radiant tube described in this paper.

The radiant tube is presented on figure 1 below.

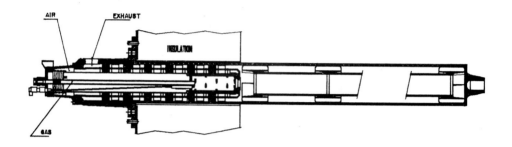

FIG. 1 : SINGLE ENDED RADIANT TUBE

2. DESCRIPTION OF THE RADIANT TUBE

The radiant tube shown in figure 1 comprises the following parts :

   2.1 The outer tube

The outer tube transmits heat to the furnace by radiation. It prevents contamination of the treatment atmosphere in the furnace by the combustion products. It must therefore be sealed along its entire length.

At present, this outer tube is made of centrifuged cast refractory steel (standard NF A : 32-057). The composition of the alloy varies slightly depending on the type of application. The most widely used steel, with 35 % nickel and 25 % chrome, corresponds to the standard NF A Z40 NC 35 25.

   2.2 The inner recirculation tube

This tube, concentric with the outer tube, provokes recirculation of a part of the combustion products. It therefore has orifices at its two extremities to allow the combustion products to pass through. In high performance single-ended tubes, this inner tube is made of ceramic (silicium carbide).

It is shown on figure 2 below.

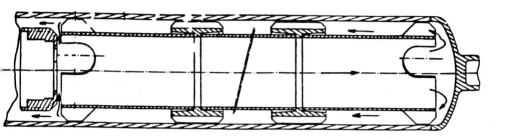

FIG. 2 : CERAMIC INNER RECIRCULATION TUBE

As no ceramic tubes of sufficient length can be produced at an acceptable price, the inner tube is made up of straight tube sections joined by ceramic connection and end pieces so that the distance between inner and outer tubes is constant along the axis, as shown in figure 2 above. The extremities of the tube touch the burner nozzle at one end and the extremity of the outer tube at the other. The connection and end pieces have a conical internal surface on both sides of the central collar to permit decentering and articulation of the inner tube.

2.3 The self-recuperative burner

This is a self-recuperative jet burner with a ceramic (silicon carbide) combustion chamber. The chamber is produced by slip casting, with rough cast tangential air inlet orifices. It is shown in figure 3. The registered brand name of the burner is CERAJET Ⓡ.

Gas is injected axially into the combustion chamber. Air enters through tangential orifices located on five places in counter rotation. The intense mixing of air and gas in the combustion chamber caused by counter rotation leads to intense combustion inside a relatively limited volume.

The mixture is ignited by means of a high voltage electrode mounted paral-

lel to the gas inlet. A high temperature glass plate closes off the rear
end of the combustion chamber.

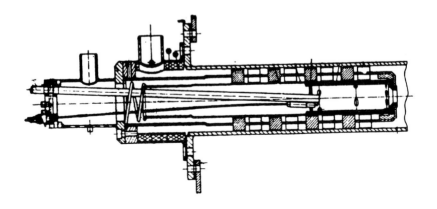

FIG. 3 THE CERAJET ®️ SELF-RECUPERATIVE JET BURNER WITH CERAMIC CHAMBER

The exchanger is made of refractory steel parts cast with internal and external fins placed back to back. The fins are staggered in order to increase heat transfer by convection. The fine fins are cast using the investment
casting technique. This fineness leads to very low air and flue gas pressure losses. It also avoids the presence of shrinkages at the foot of the
fins, as often found in parts cast using standard techniques. The parts,
each measuring 90 mm, are welded together. The end part at the burner nozzle is designed to enable the combustion chamber to rest on it.

The combustion chamber, placed inside the exchanger, is pushed against the
nozzle by means of a spring located in the cold zone against the outer
flanged mount of the burner. The ceramic combustion chamber can therefore
always operate under compression. Moreover, the dilation coefficient of the
ceramic is much smaller than that of the steel of the exchanger. Two different materials may thus react freely to thermal gradients and shocks
without creating stresses.

The flanged combustion products outlet section on the outer body of the
burner has a standard site for flue gas analysis and temperature measur-
ment. The gas inlet section comprises, in addition to the UV cell flame
viewing port mounted on the axis of the gas inlet, a standard built-in

pressure governor to measure the flowrate of inlet gas and hence the input. A viewing port is also mounted to enable the users to observe the flame. The on/off air and gas solenoid valves are placed very close to the burner in order to avoid accumulation of unburnt gases when the burner is turned on and off very frequently. A control box with high voltage transformer and safety box are built into the outer body of the burner. Each burner is thus independent. The control boxes are connected either to a control system or to a micro-computer which controls the reconfiguration of zones in a heat treatment furnace.

2.4 Regulation

The jet burner mormally operates at its nominal output, as it is designed to function with on/off control. However, variations in temperature due to the regulation cycles lead to considerable material fatigue. A new type of regulation has been developed recently which make use of self-adapting pulse modulation control systems which operate on a short time scale (32 to 64 seconds). This type of regulation, which imposes a minimum operation during this time period, avoids excessive temperature variations and hence increases the life-span of materials. The frequency of regulation cycles is increased greatly compared to standard on-off control but the amplitude of temperature variations is considerably reduced.

Figure 4 shows the effect of thermal cycles, measured by means of thermocouples inserted along the outer tube, with on/off control and pulse modulation over periods of 64, 32 and 16 seconds.

2.5 Comfort of utilization

Two principal criteria were studied with respect to comfort of utilization : noise and heat radiation from the outside of the burner.

Noise and heat radiation are limited by inner insulation of the burner with fibrous ceramic material which absorbs noise and lowers the temperature of the outer burner walls.

2.6 Maintenance

The maintenance services of heat treating plants call for burners requiring minimum maintenance time.

The CERAJET R burner was designed to satisfy this need :

- if the ignition electrode needs to be replaced, only the flanged gas inlet section needs to be dismounted. The gas inlet and the elctrode form a single element with the cold flanged section.

- if a part of the burner is broken or distorted, the entire burner can be removed by dismounting the flanged flue gas outlet section.

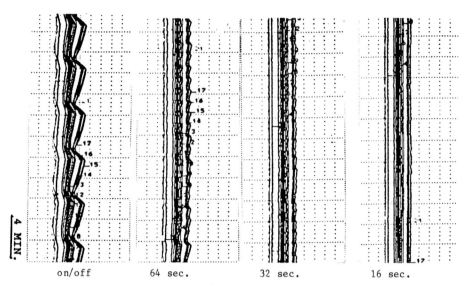

4 MIN.

on/off   64 sec.   32 sec.   16 sec.

<u>100 °C</u>

FIG. 4 : TEMPERATURE AMPLITUDE OF THE OUTER TUBE AS A FUNCTION OF
REGULATION MODES

3. TECHNICAL CHARACTERISTICS

The radiant tube described in this document was tested continuously for a period of 18 months in a GAZ DE FRANCE pilot combustion chamber. The thermal load of the installation was provided by circulation of air around a muffle protecting the radiant tube from forced convection. The aim of the test was to prove the reliability of the product and to note any anomalies encountered during the test. After 18 months of continuous operation 24 hours a day, with a regulation period of 32 seconds, no anomalies were observed either in the materials or the air/gas and ignition solenoid valves. After each month of operation, the burner was dismounted and inspected.

3.1 Operating characteristics

The radiant tube tests were conducted at chamber temperatures of 950 and
850°C, corresponding to the temperature range of carburizing and carbonitriding. The tests were conducted with LACQ natural gas with the following
characteristics :

- theoretical air requirement : 9.7 $m^3/m^3(n)$ of gas,
- net calorific value : 10.2 $kWh/m^3(n)$ of gas.

The parameters measured were the following :

- input Pe in kW (ncv),
- air factor : n,
- flue gas exhaust temperature Td in °C,
- combustion efficiency Rg in % ncv,
- heat flux Te in kW/m^2,
- % NOx represents the level of NOx emissions adjusted to 3 % oxygen in the
  flue gases.

Several operating characteristics are presented in tables 1 and 2 below.

Pe kW.(ncv)	n	Tf °C	Rg /ncv	Te kW/m^2	% NOx mg/m^3
24,4	1,1	524	75,5	20,9	625
29,6	1,1	560	73,8	24,8	573
34,4	1,1	570	73,2	28,6	540
25,4	1,2	535	72,8	21,0	593
30,5	1,2	560	71,8	24,9	520
35,3	1,2	582	70,2	28,2	499

Table 1 : results for a chamber at 950°C

Pe kW.(ncv)	n	Tf °C	Rg /ncv	Te kW/m²	% NOx mg/m³
24,9	1,1	496	76,7	21,7	535
31,1	1,1	526	75,2	26,6	495
33,6	1,1	560	73,7	28,1	489
24,6	1,2	498	74,8	20,9	522
30,5	1,2	535	73,2	25,4	477
35,0	1,2	562	71,8	28,6	435

Table 2 : results for chamber at 850°C

3.2 Analysis of results

These results show that the combustion efficiency is between 70 and 80% ncv for carburizing and carbonitriding. They were obtained on a muffled test chamber with a thermal load provided by air circulation. Measurements made on equipement installed on an industrial furnace with metal loads, revealed that combustion efficiency was raised by 3 to 5 percent under identical conditions of output and temperature. This increase in combustion efficiency can be attributed to the inertia of the load.

The temperature gradient measured along the outer tube by K type thermocouples mounted in the metal is at most 40°C.

Nitrogen oxide emissions are low for a burner with preheated air. They remain well below the German standard TA-Luft, which stipulates a maximum NOx emission level of 1100 mg./m³ for air preheated to 600°C.

The heat flux of the radiant tube is between 20 and 30 kW/m². In order to obtain higher heat flux, gas-tight ceramic outer tubes, able to withstand temperatures much greater than 1100°C must become industrially available. The single ended radiant tube presented in this paper is designed to be easily adaptable to these ceramic outer tubes when become available at reasonable prices.

4. TESTING IN INDUSTRY

In May 1989, 6 radiant tubes were installed on a conveyor carburizing furnace at the NADELLA Company in VIERZON by the SERTHEL Company, a GAZ DE FRANCE licenced manufacturer of this burner.

NADELLA is specialized in the production of bearings and journal crosses for the automobile industry. Its plants are equipped with gas-fired heat treatment furnaces. As NADELLA was facing maintenance problems with the 6 radiant tubes mounted on its conveyor carburizing furnace, the company was keen to collaborate in a pilot operation for the industrial implantation of the CERAJET 35T self-recuperative radiant tube.

The CERAJET burners are mounted horizontally on the NADELLA furnace in 2 zones of 3 tubes. They operate by "pulse modulation" in cycles of 1 minute. They have a diameter of 6 inches and a radiant length of 900 mm. The set temperatures are 900°C, in zone 1 and 890°C in zone 2. The journal cross production rate is 70 kg/h.

5. CONCLUSION

Indirect heat treatment is a field in which the investment per kWh for heating equipment is very high. It is also a sector in which advanced technologies are used (heating quality is a very important parameter). Car manufacturers, with their demanding criteria of economic viability and reliability of production tools, are particularly involved in this area.

Increased productivity requirements and strong competition from electricity in this sector pose the challenge of maintaining gas-fired heating equipment technology at the highest level. The use of high temperature ceramics to remplace refractory steel parts in the hottest parts of the radiant tube satisfies this objective, in terms of performance, reliability and maintenance requirements, while maintaining costs at a reasonable level.

Paper presented at Conf. on Ceramics in Energy Applications, Sheffield, April 1990
Session 2A

Development of ceramic radiant tube at Tokyo Gas

S. Yasuoka, I. Kikuchi, and Y. Takahashi

Tokyo Gas Co., Ltd. Industrial Customers Division
5-20 Kaigan, 1-chome, Minato-ku, Tokyo, 105 Japan

ABSTRACT: The high temperature indirect heating, at radiant tube temperatures up to 1370°C and furnace temperatures up to 1150°C has become feasible by the development of single ended ceramic radiant tube made of Si-SiC. Since NOx emitted from the metal heating furnace is limited to 180ppm in Japan, the NOx reduction technique by recirculating the exhaust gas have been applied. We have demonstrated that the NOx emission is reduced by 50% or more when the recirculating ratio of exhaust gas is 15% or more, and can clear the regulation.

1. INTRODUCTION

About 40% of all combustion furnaces constructed in Japan are indirect heating furnaces. Radiant tubes and muffles are tipical gas-fired indirect heating and heat-resistant steel is generally used for them. The furnace temperatures are limited to approximately 950°C because of material limitation. The indirect heating, such as carburizing heat treatment, bright heating and porcelain enameling etc., is operated at temperatures of 950°C or less. Therefore, the radiant tube made of conventional heat-resistant steel can be used for that indirect heating without material problems. However, the indirect heating at furnace temperatures above 1000°C is required for the heat treatment of stainless steel and high speed tool steel, powder metal sintering and so on. In these applications, the electric resistance heaters have been entirely employed because the conventional heat-resistant steel cannot cope with them. Recently, the vacuum furnace has been increasing for high quality, high temperature heat treatment fields, for example, the heat treatment of special materials such as titanium and zirconium, quenching of die steel and high speed tool steel, solution treatment of stainless steel, etc. Because of easy vacuum-holding, easy insulating as well as high temperature capability, the electric resistance heating has been also used. The gas-fired high temperature indirect heating technology is indispensable for cultivating a market of town gas for the indirect and vacuum heatings at furnace temperatures more than 1000°C. Tokyo Gas and Toshiba Ceramics have jointly developed a ceramic radiant tube (CRT). Tokyo Gas is responsible for the design of structure and dimensions of radiant tube and burner for uniform heating and the development of NOx reduction technology. Toshiba Ceramics is responsible for the development of ceramic materials. This paper presents the structure, material, performance and application of the ceramic radiant tube.

2. STRUCTURE OF RADIANT TUBE

A single-ended double tube type has been chosen from various types of radiant tubes. Since it comprises only straight tube without bend unlike U- or W-shaped radiant tube, it is easy to manufacture the radiant tube with ceramic material and attain high thermal efficiency by the self-recovery of waste heat. Owing to its structure, radiant tube of this type is simple to replace with electric heater, and moreover difficulties are small for designing furnace. More than 6000 of this type of radiant tubes were installed in the field of indirect heating less than 950°C, in past 10 years. Fig. 1 illustrates the structure of ceramic radiant tube. The radiant tube is mainly composed of recuperator, flame-holder and outer and inner tubes. Since these are connected with flanges to each other, the structure is convenient for the periodical inspection and the replacement of parts. Sectional model of radiant tube is shown in Photo 1. Burner is an extremely important component concerning the surface temperature profile of tube, production of unburnt components, turndown ratio and combustion noise level. The flame-holder is cup-shaped as shown in the figure. Gas mixed with primary air, about 30% of total air, is swirled by the burner tip. The secondary air is introduced through air inlet holes on the tapered part of the flame-holder to stabilize the flame. The rest of the air is also swirled by the cup as the tertiary air. Mixed gas is burnt inside of inner tube, then burnt gas U-turns to outside of inner tube through small holes which were provided for heat balance, and tip of radiant tube. After that, burnt gas returns through gap between outer tube and inner tube, and heat exchange is carried out with air for combustion, then exhausted out. Since the gas is mixed with the air by stages, mild and long flame without local hot spot can be obtained and the suitable combustion without the unburnt components can be

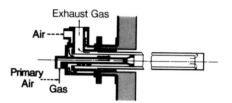

Fig. 1   Single-ended Ceramic Radiant Tube

Photo 1   Sectional model of CRT

		3in.	4in.	6in.
Outer tube length (L)		~1800	~2400	~2800
Inner tube	(a)	65	78	11 7
	(b)	53	66	102
Outer tube	(A)	90	115	164
	(B)	76	100	148

(mm)

Table 1   CRT dimensions

taken place even at an excess air ratio of 1.1 or more. This burner has good combustion characteristics enough to keep the stable combustion even by mixing with the exhaust gas to reduce the NOx emission as mentioned later. Table 1 shows the standard dimensions of ceramic radiant tube. Three different sizes, 3-, 4- and 6-inch, are available.

3. MATERIAL

Fig. 2 shows the relationship between the temperature of ceramic materials and the strength. Although the strength of ceramic material remarkably depends on the manufacturing process, the approximate values are shown in the figure to represent the features of materials. The data of heat-resistant steel is shown for reference. Non-oxide ceramic materials as Si3 N4 and SiC have excellent high-temperature strength, especially the strength of SiC does not so much decrease even at temperatures of 1000°C or higher, while that of Si3N4 decreases at those temperatures. SiC is an excellent material for using at temperatures of 1300°C or higher. SiC, as industrial material, is divided roughly into 3 types. The 3 types are Recrystallized SiC which is porous material, Si impregnated SiC (Reaction-Bonded SiC) which is high in airtightness, and sintered SiC which is the purest type. Out of these, Recrystallized SiC and Si impregnated SiC (called Si-SiC, hereafter) which have high possibility for realization at present, were selected. In addition to these two silicon nitride bonded SiC which is another representative industrial material, was taken up as test material, and trial manufacture of radiant tubes and tests have been continued with these 3 materials. Recrystallized SiC and Silicon nitride bonded SiC are low in cost but their oxidation rates are high, and were found to have problem in durability. Si-SiC has prospective durability though the cost is a

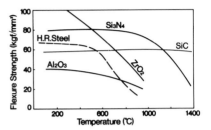

Fig. 2 Flexural Strength of Various Materials

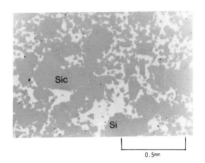

Photo 2

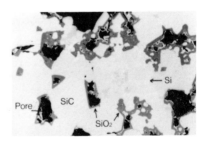

Photo 3

little higher. Therefore, it has been decided to employ Si-SiC. The oxidation rate is a key factor for determining the durability. The material is prevented from oxidizing by filling the pore of SiC with metallic Si and forming a dense film of SiO2 on the surface of it to keep away the diffusion of oxygen into the inside of material. Photo 2 is a microscopic photograph of SiC completely impregnated with Si. The white part represents Si and the dark part is SiC. Photo 3 shows an example of SiC incompletely impregnated with Si, which has cracks developed after using at 880°C for two years in a mild carburizing furnace. The black part is the pore and the dark part around it is SiO_2. Table 2 shows the physical properties and chemical analysis of the defective material. The fourth column represents the data of defective material which apparently different from these of other parts. Based on the experience, these troubles have been solved by changing the impregnation method of Si and applying the inspection by X-ray CT scanning to all impregnated ceramic tubes. Features of Si impregnated SiC which is a composite of SiC, a ceramic material, and Si, and metallic material, are summarized as follows.

• Chemical stability is high against various atmosphere.
• High in airtightness, and oxidation of material is suppressed at the surface.
• Heat capacity is low, and high in thermal conductivity.
. Bulk density is so low as 40% of heat resistant steel.
• Deformation by creep which is observed in metal is not found.
• Manufacture of product with optional shape or large sized product are possible, because sticking and processing are easy.

Table 3 shows the physical properties of Si-SiC. The data of heat-resistant steel is shown for reference. This material has an allowable maximum temperature due to the melting point of Si. Supposing allowable

		Physical Properties			Chemical Analysis	
		Bulk Density	Porosity (%)	Flexure Strength MPa	Si (%)	SiO2 (%)
Initial Data		3.00	0.1	180	16.0	0.3
Outer Tube	①	3.02	0.06	150	11.6	5.2
	②	3.02	0.02	200	15.2	2.2
	③	3.01	0.02	170	16.1	2.2
Inner Tube	④	2.86	0.73	90	5.7	16.5
	⑤	3.02	0.06	200	17.0	2.0
	⑥	3.01	0.03	180	14.7	2.5

Table 2   Physical properties & chemical analysis after long run

Physical properties \ Materials	Si-SiC	Heat resistant steel (35Cr-25Ni)
Bulk density (g/cm3)	3.0	8.05
Modulus of repture (kgf/cm2)	2040	—
Thermal shock resistance ΔTc (°C)	400~450	—
Thermal expansion (at 1000°C)×10⁻⁶	4.2	18.4
Specific heat (kcal/kg°C)	0.17	0.11
Thermal conductivity (at 1000°C) (kcal/mh°C)	57.8	1.83
Max. usable temperature (°C)	1370	1100

Table 3   Physical properties

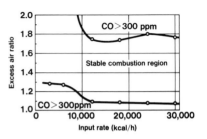

Fig. 3   Stable Combustion Region

temperature of this material as 1370°C, then maximum temperature in furnace will be about 1100 to 1150°C.

4. PERFORMANCE

4.1 Combustibility

Fig. 3 shows the stable combustion region of 3 inch ceramic radiant tube in open air. For the purpose of using in high temperature, opening of burner tip of customary metallic type was made wider, but wide stable combustion region similar to ordinary burner, was obtained. For the limit of low load, it is necessary to make heat flux of tube higher than 1.0Kcal/cm^{2}h, by the reason to suppress the rise of temperature of flame-holder and generation of soot which is caused, by incomplete mixing. Moreover, for the limit of high load side, furnace temperature and heat flux shall be kept lower than 1150°C and 2.7Kcal/cm^{2}h respectively. Combustion noise is about 65dB(A) when input rate is lower than nominal rate.

4.2 Tube temperature profiles

Fig. 4 shows the surface temperature profiles of a 3 inch 1000mm long radiant tube at the furnace temperature of 1100°C and the heat flux of 4Kcal/cm^{2}h. Similar to usual metallic type burner, small holes were provided on inner tube in order to keep heat balance. The figure shows that the temperature profiles are almost uniform and the temperatures of inner tube are approximately 100°C higher than these of outer tube.

4.3 Preheated air and exhaust gas temperatures

Fig. 5 shows the preheated air temperature and the exhaust gas temperature at input rate of 6000 and 13000Kcal/h. It is a matter of course that these temperatures depend on the furnace temperature. For instance,

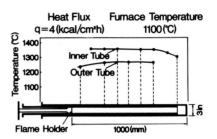

Fig. 4   Temperature Profiles

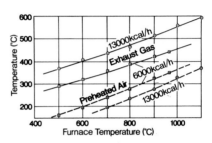

Fig. 5   Air & Exhaust Gas
Temperature

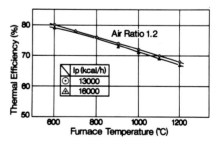

Fig. 6   Thermal Efficiency

the preheated air temperature and the exhaust gas temperature are approximately 350 and 500°C, respectively, when the furnace temperature is 1000°C. The lower the input rates, the higher the preheated air temperature is and the lower the exhaust gas temperature is, because of the recuperator characteristics.

4.4 Efficiency

Fig. 6 illustrates the thermal efficiency with the furnace temperature, where the efficiency is defined as the efficiency calculated regarding only the loss by the exhaust gas as the heat loss. The figure shows that the thermal efficiency at the furnace temperature of 1000°C is more than 70%.

4.5 NOx

Fig. 7 shows the NOx emission calculated on the basis of oxygen concentration of 11%. Since NOx emitted from the metal heating furnace is limited to 180ppm according to the NOx regulation in Japan, the NOx emission exceeds the regulation level when the furnace temperature is 950°C or higher. We have applied the most reliable, less-risky NOx reduction technique shown in Fig. 8 that the NOx emission is reduced by recirculating the cooled exhaust gas. The exhaust gas is cooled down to 80°C in a water-cooling heat exchanger and the cooled exhaust gas is mixed with fresh air by a blower. Fig. 9 shows the effect of the cooled exhaust gas on the NOx reduction at the furnace temperature of 1100°C. The figure shows that NOx is reduced by 50% or more when the recirculating ratio of exhaust gas is 15% or more. Fig. 10 shows that the thermal efficiency is reduced from 70 to 65% at a recirculating ratio of exhaust gas of 20%. From these results, we can clear the NOx regulation with a small decrease of the thermal efficiency. Now, we are going to

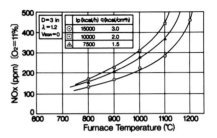

Fig. 7 NOx Emission

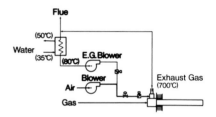

Fig. 8 Exhaust Gas Recirculation Method

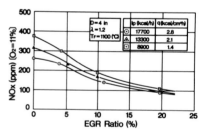

Fig. 9 NOx Reduction by EGR

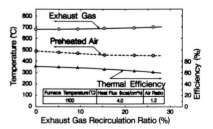

Fig. 10 EGR and Thermal Efficiency

develop a new exhaust recirculator using inlet air momentum to reduce the Nox without lowering the thermal efficiency. It will be completed shortly.

5. APPLICATION

Development of ceramic radiant tube was started in May 1985, and 7 furnaces and 153 tubes were delivered actually. Examples of application of ceramic radiant tube are as follows. Photo 4 shows a batch type carburizing furnace using 8 ceramic radiant tubes. Small bolts for automobile were mild carbo-quenched, and 4 years passed after the start of operation with furnace temperature of 880°C.

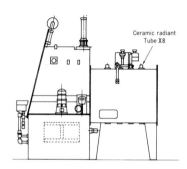

Photo 4   Batch type carburizing furnace

Photo 5 shows another furnace at the same customer. It is a mesh belt type continuous non-oxidizing quenching furnace. This furnace uses 6 ceramic radiant tubes among 18 single-ended radiant tubes. The furnace temperature is also 880°C and these tubes have been used for 3 years.

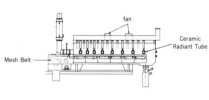

Photo 5   Continuous non-oxidizing furnace

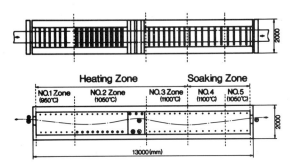

Heating Zone			Soaking Zone	
NO.1 Zone (950°C)	NO.2 Zone (1050°C)	NO.3 Zone (1100°C)	NO.4 (1100°C)	NO.5 (1050°C)

13000(mm)

Fig. 11   Catenary Type Continuous
Annealing Furnace

Photo 6 is a newly built catenary type continuous annealing furnace for stainless steel strip. Furnace temperature is 900 to 1150°C, and input rate is 1,100,000Kcal/h max. 73 ceramic radiant tubes in total, 60 3-inch tubes and 13 4-inch tubes, are installed. Fig. 11 shows the installation of radiant tube, and the furnace with total length of 13m is divided into 5 zones, the temperatures of which are controlled respectively. Photo 7 is the heat exchanger for exhaust gas recirculation (EGR) which was equipped for the reduction of NOx. Fig. 12 shows the result of measurement of NOx reduction effect and the result shows good agreement with experimental data. In this furnace, EGR ratio was set at 15%. In the present time, 18 months has passed after installation, troubles concerning material of CRT or troubles on combustibility were not observed. From the results of check of sample, the progress of oxidation, etc. were found to be kept low, hence, life longer than initial expectation is anticipated. Photo 8 shows vacuum furnace. Checks on sample of CRT outer tube were conducted after 130 hours operation which include 5 times treatment at furnace temperature of 1060 to 1100°C and 10^{-2} to 10^{-3} Torr, but deterioration by vaporization by Si, etc. was not observed absolutely. Airtightness is kept by applying O-ring to flange of outer tube and water cooling.

Photo 6   Catenary type continuous
annealing furnace

Photo 7   Heat exchanger for EGR

6. CONCLUSIONS

As mentioned above, the indirect
heating at tube temperatures up to
1370°C and furnace temperatures up to
1150°C has become feasible by the
development of single-ended ceramic
radiant tube made of Si-SiC. The
attainment of the goal to reduce the
NOx emission has been demonstrated by
the exhaust gas recirculation.
Careful attention for setting of CRT,
because it is sensitive to mechanical
shock, and avoidance of thermal shock
such as pouring water at the
condition of elevated temperature
(cooling by air for combustion during
off time, has no problem), etc. are
important conditions. In addition,
some problems are remaining in
handling and maintenance including
durabilities of auxiliary equipment
such as spark rod at high
temperature. However the ceramic
radiant tube has the following
advantages; it can be used at high
temperatures than conventional
heat-resistant steel radiant tube;
the strength is not lowered even at
high temperatures; the tube supporter
can be eliminated; the ceramic
radiant tube is hardly deformed; and
the corrosion resistance is
excellent. From these advantages, the
ceramic radiant tube may become to be
widely used not only at high
temperatures but also at temperatures
below 1000°C instead of the
heat-resistant steel radiant tube.
Moreover, Si-SiC which has various
excellent performances, is expected
to has wide field of application such
as muffle for external heating,
element and unit for pipe or plate
for high temperature heat exchanger.
Photo 9 shows a vacuum rotary furnace
for baking super conductor. This
furnace, using Si-SiC tube as muffle,
provides uniform heating and
promotion of reaction of contained
materials at 1250°C max and 10^{-2} to
10^{-3} Torr, and operating
satisfactorily since the installation
in June 1989. Taking these

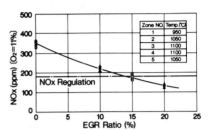

Fig. 12   NOx Reduction in Catenary
          Furnace

Photo 8   Vacuum furnace

Photo 9   Vacuum rotary furnace

application in other field into consideration, confirmations of durabilities of CRT in various condition are scheduled in the future. Further, CRT of Sintered SiC which is excellent in heat resistance, homogeneity and air-tightness, is under development, and application to the use in higher temperature is expected possible in the near future.

An overview of the UK National Programme for Ceramic Applications in Reciprocating Engines (CARE)

D A Parker, BSc, PhD., ARCS, CEng, CPhys, FIMechE, FInstP.
Chairman, CARE Consortium
Managing Director, T&N Technology Ltd., Cawston House, Rugby, CV22 7SA.

ABSTRACT:

The paper traces the origins of the CARE programme against a background both of early experimentation with ceramic engine components and of enhanced awareness of the potential role of new and improved materials and processes. It describes the work of the Consortium - a collaborative venture between 30 organisations participating in 12 projects - under its four main Technical Sections.

Within the Insulation Section, projects have ranged from fundamental studies of the effect of insulation on combustion to assessment of the improvement in vehicle fuel economy due to use of a smaller cooling system.

The Substitution Section has investigated the potential benefits derived from better wear properties (tappets and rockers), lower mass (poppet valves) and the possibility of reduced emissions from ceramic pre-combustion chambers.

The Turbocharger Section has considered two very different ceramic applications, namely to the rotor of an automotive turbocharger and to the bearings of a much larger machine.

The three user driven sections have been complemented by a Materials Section seeking to provide improved materials and processes to satisfy current and projected user requirements. Its innovations have included a potentially cheaper route to zirconia powder.

The paper reviews the major results achieved, particularly in terms of their potential impact on future engine technology.

1. INTRODUCTION

Exploration of the use of ceramics as structural components in reciprocating engines began around 1970 (1) when use of a reaction bonded silicon nitride (RBSN) piston was demonstrated in a Villiers 0.9 kW, 50 mm bore gasoline engine. Work on diesel engines began with a 108 mm diameter piston for a single cylinder Gardner 9 kW engine (2) in which significant running was achieved (and surprising user-friendliness demonstrated inasmuch as a templug that had worked loose resulted in a neat groove in the piston rather than the expected large scale fracture). In a further and more detailed investigation (3) a highly rated 15 kW 80 mm diameter

Petter diesel piston demonstrated a 50% survival rate after extensive testing, using monolithic ceramic construction or mechanically attached ceramic crown pieces. At about this time work in the United States (4) predicted significant gains in economy due to combustion chamber insulation followed by turbo compounding to recover the additional power diverted to the exhaust system. In these early days combustion chamber insulation was seen as providing a major potential market for ceramics, largely due to progression of the American work to the point of demonstration of the on-highway operation of a military truck with a completely uncooled engine (5).

During the 1980's work on potential applications of ceramics in reciprocating engines has broadened to include those which arise from the properties of low wear rate and low inertia as well as low thermal conductivity. The major components so considered are listed in Table 1.

INSULATION	LOW WEAR RATE	LOW INERTIA
Glow Plug Pre-Combustion Chamber Piston Cylinder Liner Cylinder Head Plate Exhaust Port Liner Turbocharger Casing	Rocker Arm Tip Tappet Bucket Disc Valve Guide Valve Seat Roller Follower - Cam Piston Ring Fuel Injector	Turbocharger Rotor Poppet Valve Piston Pin

TABLE 1
POTENTIAL APPLICATIONS OF CERAMICS IN RECIPROCATING ENGINES

Most of the work has concerned monolithic ceramic components, but ceramic powders have been applied in the form of sprayed coatings to provide increased thermal insulation, increased wear resistance and increased corrosion resistance. Typical monolithic components are illustrated in Figs. 1-3.

By the mid 1980's there was a general realisation that many, if not most, of the post-war innovations had been materials based - the transistor is perhaps the best example. In consequence, H.M. Government appointed what has become known as the Collyear Committee (6) to consider the significance of recent materials developments and to prepare a programme for the wider application of new and improved materials and processes. In their report the Committee emphasised the total interdependence of material and process, meaning that in the great majority of manufacturing processes the end result is dependent both on the starting material and its subsequent history. It was thus clear that in any significant application of new materials, such as ceramics, the processing would need to be as carefully considered as the raw material itself. In their report the Committee listed seven main categories of materials as worthy of further systematic development, namely:-

1. Composite Materials
2. Engineering Ceramics
3. Rapid Solidification Technology
4. Electronic Materials
5. Near Net Shape Manufacture
6. Surface and Joining Technology
7. Assurance of Product Performance

FIG. 1
CERAMIC VALVE TRAIN PARTS
AND PRE-COMBUSTION CHAMBERS

FIG. 2
CERAMIC INSERTS USED FOR PISTON,
LINER AND EXHAUST PORT INSULATION

FIG. 3
CERAMIC TURBOCHARGERS

The inclusion of engineering ceramics in this prestigious list significantly enhanced the prospect of setting up a collaborative research programme with financial assistance from the Dept. of Trade and Industry. Realising that our major foreign competitors, in the United States, Japan and Germany were already engaged in collaborative work in ceramics and that there was in the U.K. a significant base of materials expertise and early applications work on which to build, the first offer letters for the present programme were issued by the Department of Trade and Industry early in 1986. This collaborative research programme on Ceramic Applications in Reciprocating Engines (CARE) was to run for four years, ending in December 1989. Currently some 30 organisations are participating (Table 2) in 12 clearly defined projects.

British Ceramics Research Ltd. Napier Turbochargers Ltd.
BP Research Centre NEI-APE Ltd.
Castrol Ltd. Noel Penny Turbines Ltd.
Cookson Group plc Perkins Engines Group Ltd.
Dynamic-Ceramic Ltd. Pilkington Bros plc
Esso Petroleum Co. Ltd. Ricardo Consulting Engineers plc
Exxon Chemical Ltd. Rolls-Royce Ltd.
Fairey Tecramics Ltd. Schwitzer Household Manufacturing Ltd
Ford Motor Co. Ltd. Shell Research Ltd.
Harwell Laboratory GR Stein Refractories Ltd.
Jaguar Cars Ltd. T&N Technology Ltd.
Leyland DAF Ltd. Universal Abrasives Ltd.
Lister-Petter Ltd. University of Bath
Lucas Diesel Systems Vesuvius Zyalons Midlands Ltd.
Morgan Matroc Ltd. Wellworthy Ltd.

TABLE 2
ORGANISATIONS PARTICIPATING IN THE CARE PROGRAMME

In considering the form of organisation for the CARE Consortium, the Management Committee took full account of the conflict between exaggerated market predictions based upon an extrapolation from the early results and the hard competitive reality of the engine components market. In almost every market (Japan is a possible exception) ceramics would be accepted in reciprocating engines only if they were able to provide some aspect of performance not currently available or to provide a currently available aspect more competitively. Moreover, the degree of difference would have to be large enough to justify the cost of replacing existing manufacturing facilities. In considering cost-effectiveness, however, it is accepted by the engine manufacturers that one can take a 'systems' view. Thus a particular component manufactured in ceramic rather than metal may currently be more expensive, but may provide advantages that have a commensurately greater cost benefit. Examples might include combustion chamber components that contributed to reduced emission of particulates, or valve gear components whose reduced inertia allowed the valve train, and hence engine, specification to be uprated.

With these considerations in mind, it was decided to emphasise the applications aspect of the programme by making three of the Consortium sections user-lead. These sections are concerned with thermal insulation, materials substitution and turbocharging. The fourth section, concentrating on materials development, has been guided by the materials requirements of the user sections, but has also been encouraged to use its expertise and innovation in developing novel materials and processes for current and future application. The remainder of the paper will now consider the work of the four Consortium sections, subdivided to encompass the 12 technical projects.

2.   INSULATION SECTION

2.1   Thermally Insulated Diesel Engines

This project was one of the first to start and is being coordinated by Leyland DAF. The objective was to construct, and undertake detailed performance measurements on, a production diesel engine whose combustion chambers had been modified by thermal insulation. Phase 1 of the programme was effectively a reconstitution of the insulated engine (7) prepared by Leyland some years earlier. Adoption of this approach provided a well documented baseline for further development and a rapid start to the engine running aspect of the project.   Early tests demonstrated that the previously attained insulation level (33% reduction in rejection of heat to coolant) had been regained.   To take advantage of this reduction the engine cooling system was redesigned and tested, and the insulated engine tested both on the test bed and in a vehicle.   Whilst no fuel economy improvements were apparent on the test bed the smaller cooling system allowed an improvement in fuel economy to be realised in the vehicle (8).

Under Phase 2 of the project the combustion chambers have been insulated to a higher degree using a number of alternative monolithic components and coating techniques.   The original insulated pistons were essentially standard Leyland pistons, machined back and plasma sprayed with about 1.5 mm thickness of zirconia.   By May 1989 these had achieved around 200 hours durability.   The next piston to be tested was the ceramic sprayed, metal crown design depicted in Fig. 4 in which the ceramic coating served both to increase the degree of insulation and to provide thermal protection of the underlying superalloy.   The ceramic crown piece design, also shown in Fig. 4, has achieved around 150 hours durability in other engine testing.   Retention of the reaction bonded silicon nitride crown is achieved by shrinking into a superalloy sleeve.   In this design, too, insulation is achieved partly by ceramic and partly by the underlying air gaps.   However, problems of coating detachment are avoided by use of the monolithic ceramic crown piece.

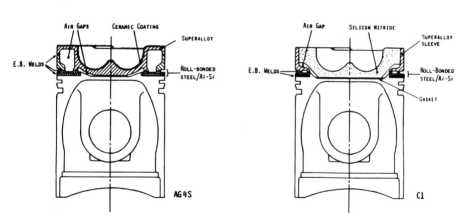

FIG. 4
AIR GAP AND CERAMIC INSULATED PISTONS FROM T&N TECHNOLOGY

Other monolithic components included in the programme are zirconia cylinder liners, restricted in length to lie above the point of top ring reversal (Fig. 2), and zirconia cylinder head plates.

Results of the Phase 2 tests are encouraging, but require further evaluation.   However, research going on elsewhere (9) suggests that the most effective in-cylinder components to insulate are the cylinder head plate and the upper parts of the liner.   Other aspects of the current test programme have included investigation of the effect of fuel cetane number and the evaluation of a specially developed high temperature lubricant.

2.2 Fundamental Rig and Engine Trials of Ceramic Coated and Monolithic Components

As its name implies, this project was introduced to facilitate the development of novel insulating materials for diesel pistons without incurring the heavy costs associated with multi-cylinder engine testing. Within this framework it was planned to evaluate materials first in a specially constructed test rig and then to proceed with the most promising to a single cylinder engine. A major feature of the test rig was the achievement of heat fluxes comparable with those expected within the engine. By careful disposition of infra-red lamps the heat flux has been built up to 450-500 kw/m^2 and the rig is now performing extremely well.

Engine testing is now being performed in a Petter PH1W engine, converted to indirect injection to allow read-across of the results to automobile diesel practice. Piston crown insulating means so far tested include Cordierite and Lithium Aluminium Silicate (LAS) crown pieces as well as Partially Stabilised Zirconia (PSZ) plasma sprayed onto an aluminium substrate. The test rig has shown itself capable of creating failures by thermal cracking and thus of separating out those materials unlikely to survive in a firing engine. Up to May 1989 only 2 out of 9 Cordierite discs had failed in the thermal shock rig, although there were indications that the test conditions were even more severe than those encountered in an engine. Cordierite piston caps have been supplied for the engine test and LAS caps are expected shortly. A subsidiary investigation has shown the importance of surface finish and has indicated that strengths in excess of 200 mPa can be achieved in Cordierite with well finished samples.

2.3 Ceramic Exhaust Port Liners (Fig. 2)

Reduction of the heat transfer from exhaust gases in the port region to the surrounding parts of the cylinder head has many advantages, particularly in an engine with an insulated combustion chamber. These include lower heat transfer to coolant, lower thermal stresses in the cylinder head, lower exhaust pipe external (and hence lower underbonnet) temperatures, and higher exhaust gas temperatures with their greater capacity to operate turbo machinery or exhaust treatment devices in the exhaust pipe. The turbo machinery might be used for a number of purposes including the conventional pressurisation of incoming air, but might eventually include feedback of mechanical power to the crankshaft, generation of electrical energy required elsewhere in the system or for other purposes.

Work to develop suitable port liners has concentrated in the use of aluminium titanate, first of all in the form of simple tubes and later in slip cast shapes to be subsequently incorporated in a cast iron cylinder head. Early work on 3 mm walled tubes indicated a problem of cracking or spalling in the ceramic and high residual tensile stresses in the surrounding cast iron. The next phase in the development was to devise and apply coatings to attenuate the casting stresses (which arose due to both thermal shock and the difference in expansion between the aluminium titanate and the surrounding iron). Further casting trials showed that with 3.5 and 5 mm wall thickness tubes coatings were useful in minimising cracking, though prevention could not be assured. Problems associated with the complicated shape of the port liner were overcome by fettling the slip cast liner. The next step will be the manufacture of cylinder head castings incorporating these port liners for a Perkins single cylinder engine.

2.4 High Temperature Combustion in DI Diesel Engines

The contribution of this project to the overall CARE portfolio has been to provide a better understanding of the effects of combustion chamber insulation on the in-cylinder processes. In early work (3) it was realised that the nature of the combustion had been changed but, to make best use of the material and component developments elsewhere in the programme, more quantitative information was required on the changes to combustion, ignition and fuel emissions and fuel tolerance. The project also afforded the opportunity to reoptimise an engine once insulated. Since the baseline (non-insulated) engine would have been optimised by a century of development, it was only reasonable that the insulated engine with which it was compared should also be optimised. So far this has been attempted in very few of the published studies around the world. In a thorough investigation, using their Proteus engine, Ricardo began by evaluating the baseline and insulated baseline builds. The third phase of the project was to optimise the insulated build and finally to evaluate its characteristics.

The poor combustion characteristics of the insulated baseline build were substantially improved by the addition of high pressure fuel injection equipment together with a nozzle optimisation study.   The improved combustion characteristics resulted in decreased particulate emissions but increased NOx emissions. The optimised build also exhibited a circa 3% improvement in fuel economy compared with the baseline build at the rated power condition at optimum timing.   Overall reductions of 20-25% in the amount of fuel energy transferred to coolant were also achieved by the optimised build.

European 13 mode emissions calculations showed that substantial injection timing retard would be necessary to reduce NOx emissions to realistic levels.   This severely penalised both particulate emissions and fuel consumption, compared with the normally cooled baseline engine.

Piston temperature measurements revealed peak temperatures above 800°C at full load. However, piston temperature responses to changes in speed and load were not as expected, with highest temperatures observed at the low speed and peak torque conditions. The temperatures were measured using thermistors and a 2-beam mechanical linkage. The values recorded at 1300 rev/min, 16.8 bar bmep for the ceramic crown piece piston illustrated in Fig. 4 are presented in Fig. 5.   Other results show that the piston centre (the 'pip') was hottest at 1000 rev/min, 12 bar bmep and became cooler as the speed increased. The edge of the bowl was hottest at peak torque (1300 rev/min) and again became cooler as the speed increased. In an extended programme, but still within the current CARE time frame, the facilities set up for the project are being used to investigate alternative methods of NOx reduction and the magnitude of instantaneous heat transfer characteristics in insulated engines, which has recently caused much debate on the international scene (10).

3.   SUBSTITUTION SECTION

3.1 Basic Tribology

Early experience with insulated cylinder components quickly revealed that significantly higher temperatures than normal would result at the sliding surfaces. As the low wear rates attainable with ceramics became more widely appreciated, quantitative data were required for load/speed

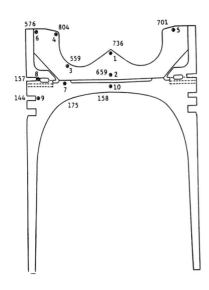

FIG. 5
OPERATING TEMPERATURES (°C) OF CERAMIC
CROWN PIECE PISTON IN RICARDO PROTEUS
ENGINE AT 1300 REV/MIN, 16.8 BAR BMEP

combinations characteristic of valve train operation. In view of conflicting experimental evidence, and the need to build up a sound background of basic tribology on which to plan future applications, this project was instituted to determine the friction and wear properties of various ceramic materials (usually but not necessarily against metals) as a function of surface finish, load, speed, temperature and degree of lubrication.

To represent in-cylinder applications, reciprocating sliding wear tests have been undertaken on tetragonal partially stabilised zirconia (TZP) components. Test conditions varied from copious lubrication at 300°C to drip feeding at 140°C, though it was interesting to note that over this range of conditions the levels of wear and friction were related to the quality and fineness of the surface finish.

Under conditions representative of valve train operation, silicon nitride and sialon cams have been compared in an instrumented VW Golf 1.6 litre gasoline engine. The beneficial effects of additional spray-on lubrication, namely reduction of cam lobe temperature with consequently reduced friction, showed that boundary lubrication conditions were present. The higher operating temperature of ceramic components, arising from their lower conductivity, tends to induce boundary lubrication by reducing the viscosity of the lubricant.

3.2   Insulated Indirect Injection Pre-Chamber

Indirect injection (IDI) pre-chambers are conventionally made from nimonic alloy steel, an intrinsically expensive material. The purpose of this project was to evaluate the potential of ceramic pre-chambers to substitute for the nimonic material, bearing in mind that the lower conductivity of the ceramic would be expected to cause higher operating temperature and hence potentially reduced hydrocarbon emissions.

In Phase 1 of the project only the lower part of the pre-combustion chamber was fabricated from ceramic. The latter was assembled into the engine by shrinking into a metal supporting sleeve. Four alternative materials, namely sialon, reaction bonded silicon nitride (RBSN), aluminium titanate and partially stabilised zirconia (TZP) were evaluated, beginning with low speed and load conditions. Comparison of the engine running and thermal shock tests led to the conclusion that while some remained as potential materials, others disintegrated under high temperatures or were prone to thermal cracking and were consequently rejected. Interestingly, no significant deterioration in specific fuel consumption or increase in NOx level was observed.

The surviving materials are currently being investigated under Phase 2 of the programme in which both upper and lower parts of the pre-chamber are constructed from monolithic ceramic.

3.3 Lightweight Valve Gear

The main objectives of this project, led by Jaguar Cars, were to reduce the weight and improve the durability of selected valve train components. The first component was the tappet insert shown just above the bottom right hand corner of Fig. 1. Three monolithic materials were tested, namely sialon and two versions of sintered silicon nitride. The load and speed of operation of the tappet insert, operated against the standard steel cam, was varied over the entire operating range of the engine. The highest wear rates occurred when low speeds were combined with high loads. However, over a range of conditions the tappet wear was reduced by a factor of 10 or more relative to the standard material; the cam wear was more variable, but the best improvements were as great as for the tappet. Further friction and wear tests are to be conducted on the three selected insert materials.

The saving of weight in valve gear intended to operate at high engine speeds is important, not so much for the weight saving itself, but for the reduction of force needed to provide the necessary acceleration and the consequently reduced friction. The benefits may be taken in various ways, for example by operating with greater margins of safety, reducing the size of some driving components or - most probably - increasing the permitted valve train accelerations and thereby engine performance.

In an enhancement of the project initiated in late 1988, complete bucket tappets were manufactured in ceramic material to allow direct replacement of the steel tappets in the Jaguar AJ6 engine. The ceramic tappets are illustrated in Fig. 6. The first tappets of this type were made by machining from a blank and will be used to determine the effect of wall thickness and surface finish. Preliminary results have shown good durability and low wear. Manufacture of such shapes by injection moulding has been shown to be feasible and tests on such tappets will follow closely behind those on the machined samples.

Another component being investigated under this project was the poppet valve itself, illustrated in the top left corner of Fig. 1. Although the head of the valve is an orthodox shape, the design of the tip to receive the collets through which the valve is connected to its spring has been optimised for manufacture in ceramic. Under motoring conditions without the application of cylinder pressure, sialon and sintered silicon nitride valves ran satisfactorily. However, the application of cylinder pressure led to four valve failures, probably due to the generation of tensile stresses in the lower part of the valve head due to application of gas pressure to its working face. In consequence, a test facility has been manufactured which will provide cyclic pressure of up to 75 bars at 125 Hz frequency. The baseline conditions are being established with steel valves before tests on the ceramic valves proceed. Tensile loading of the valve stem does not seem to be a problem, inasmuch as three ceramic valves have been tested in a special proof loading rig and have achieved adequate tensile load performance.

FIG. 6
MONOLITHIC CERAMIC BUCKET TAPPETS

3.4 Dry Top End Valve Gear

All metal valve trains are copiously lubricated to prevent failures by
pitting or scuffing at the heavily loaded cam/tappet interface. This oil
also serves as a heat transfer medium, cooling the lubricated components.

In view of the good high temperature properties of ceramics and the low
wear rates established under some operating conditions, their use offered
the attractive prospect of a valve train without the need for
lubrication. Apart from reducing oil pumping losses, this would permit
the use of a smaller oil pump for the other components and eliminate the
expensive provision of drilled oil holes.

In project work conducted both at Perkins and Lister-Petter, ceramic valve
train components, including valve guides, push rod ends, rocker tips and
bushes (see Fig. 1) have been tested. The ceramic materials have included
silicon carbide, sialon, silicon nitride, partially stabilised zirconia
(PSZ) and zirconia toughened alumina (ZTA). Work began under totally dry
conditions but although some ceramic components survived limited testing,
it was concluded that such conditions were too severe for the components
and some limited lubrication would be advantageous. Despite the redesign
of push rods and adjusting screws to minimise contact stresses, the
testing of dry ceramic components continued to produce high wear, as did
the testing of standard metallic parts under the same conditions. Some
limited work using dry lubricants showed that whilst there was some
benefit, much work outside the scope of the project would be required to
develop such a system.

Accordingly, an engine-based test rig was built at Perkins with minimal oil applied to the sliding surfaces during assembly. In sharp distinction to the totally dry experience, various ceramic components with this incidental lubrication, augmented by 2/3 drops per day, were run at 1300 rpm for a total of 500 hours. During this period the wear of sialon, silicon nitride and PSZ materials was minimal. In the final stages of the project at Perkins, the most promising ceramic components from the rig tests will be progressively introduced into an actual engine to establish durability and wear performance.

The work at Lister-Petter, carefully co-ordinated with that at Perkins, gave very similar results. Engine running using only 'build lubrication' and oil mist from the crankcase was most encouraging. Engine testing at Lister-Petter continues.

4.   TURBOCHARGER SECTION

4.1 Large Turbocharger

Work on large turbochargers began with consideration of a number of alternative components that might with advantage be manufactured in ceramic. However, the list has gradually been refined to thrust and journal bearings in several sizes of machine and finally to the components illustrated in Fig. 7, a simplified version of the bearing currently manufactured in lead-bronze.

FIG. 7
RBSN BUSHES FOR LARGE ENGINE TURBOCHARGER

Axial thrust is resisted by a ring containing a number of radial oil entry regions followed by an inclined face. The ceramic version of this ring is bonded to the bearing housing assembly. The component on the right of the picture is a lobed journal bush which can either be fixed or allowed to float in the radial direction to provide additional squeeze film damping. The overall design of the bearing provides for a high thrust capacity combined with good stability and minimum power loss. Both bearing components have been supplied in sialon, sintered silicon nitride and carbide and reaction bonded silicon nitride and carbide. Testing is well advanced and has been most encouraging in that the low wear rates obtained suggest that the ceramic might provide a useful means of combating wear and corrosion. Such wear arises from fine particulate matter in the oil during start-up and shutdown, and the corrosion from sulphur in the residual fuel frequently used in the large engines fitted with such turbochargers.

4.2 Vehicle Turbocharger

Work on this project began using flat bladed turbocharger rotors as a simple means of screening alternative ceramic materials without incurring the expense of providing truly three dimensional shapes in each material. The approach also meant that spin testing procedures could be refined without undue concern for the component tested. Furthermore, comparison could be made between experimental results and relatively simple mathematical models of the component.

Using this approach rotors of sialon, silicon carbide and silicon nitride have been tested. Speeds of over 150,000 rev/min can readily be obtained providing the assembly is balanced to typical turbocharger limits.

In the meantime, an appropriate design of curved bladed rotor was identified and appropriate manufacturing means developed. The first samples were machined from the solid, but good progress with the development of injection moulding allowed the rotors illustrated in Fig. 3 to be manufactured in fully dense silicon nitride and those in Fig. 8 to be manufactured from reaction bonded silicon carbide. Testing of the curved bladed rotors is now underway.

FIG. 8
REACTION BONDED SILICON CARBIDE TURBOCHARGER ROTORS
MANUFACTURED BY INJECTION MOULDING

5.   MATERIALS SECTION

5.1 Improved Zirconia

The primary objective of this project was to establish a competitive U.K. manufacturing capability for electro-fused zirconia powders capable of manufacture into monolithic ceramic components and of application as sprayed ceramic coatings. Electro-refined zirconia powders have indeed been developed (11), with sinter activities comparable to those of chemically co-precipitated materials. These electro-refined PSZ powders can be used to produce dense engineering ceramics with good mechanical properties. The ceria-PSZ system in particular produces extremely tough monolithic ceramics with high Weibull Moduli. By controlling the chemistry of this system, particle size and firing schedules, the ceramic can be engineered to give optimised properties in a manner analogous to metal alloy property control. Typical mechanical properties are illustrated in Fig. 9.

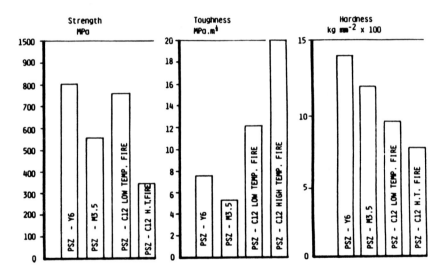

FIG. 9
TYPICAL MECHANICAL PROPERTIES OBTAINABLE FROM
ELECTRO-FUSED ZIRCONIA POWDER

The novel powders developed under this project have been shown to be comparable to other powders in spray characteristics, but offer benefits because of their chemical consistency and homogeneous crystal structure. The comminution and classification techniques developed are capable of giving the closely graded particle size required for a good plasma spray powder. The application of such coatings is also being evaluated under the CARE programme, with particular reference to the exhaust valve of a medium speed diesel engine. Preliminary assessment of alternative coating formulations was made on the exhaust valve of a Petter single cylinder diesel engine.

The most promising coatings from this trial have now been applied to the flame face of exhaust valves from a medium speed diesel engine at NEI-APE. Valve operating temperature and corrosion measurements will be made. If successful, this thermal barrier coating will allow the maximum rating of the engine to be raised without incurring a penalty due to corrosion from sulphur in the residual fuel.

5.2 Novel Materials for Thermal Barrier Coatings

Although the material discussed in Section 5.1 was manufactured by a novel route, its chemical composition was unaltered. The objectives of this project on novel thermal barrier coatings was to design the crystal structure of non-zirconia coatings so as to achieve the required properties. These were judged to be a thermal expansion equal to or greater than tetragonal zirconia at 1000°C ($10x10^{-6}$/K), a thermal conductivity within a factor of 2 of that of tetragonal zirconia (2 W/mK) and a melting point above 1500°C. In the search for such a material a large number of compound oxides were evaluated, of types ranging from the relatively simple ABO_4 to compounds as complex as $A_2B_2O_7$. A and B were selected from the series Ce, Ba, Ca, Mg, Si, Al, Zr and Ti; other more complicated oxides with three metallic radicals were also being evaluated. The preliminary results showed encouraging potential for meeting the criteria defined above.

6. DISCUSSIONS AND CONCLUSIONS

In preparing this overview it has not been possible to give the latest results from each of the projects, partly because the Consortium work continues until the end of December 1989 and partly because the financial contributors to the CARE programme need time to digest, assimilate and apply the results themselves. However, even an interim view must reflect a number of important findings that have already been established, namely:

6.1 Combustion chamber insulation can be used to divert significant heat away from liquid cooling systems to the exhaust, allowing a reduction in cooling system volume with consequent packaging benefits in a vehicle. Thus at a stage of development in which there was no fuel economy gain on the engine test bench, significant gains were realised in a vehicle.

6.2 The first ceramic components tested were direct substitutes for the corresponding metal component, i.e. there was no reoptimisation to take into account the very different properties of the ceramic. This reoptimisation has now begun, for example, in the combustion chamber described in 2.4 and the valve stem in 3.3. Reoptimisation of the combustion chamber, for example, transformed a deterioration in fuel economy to a 3% gain combined with diversion of 20-25% of fuel energy from coolant to exhaust. It has also been possible to define fuel economy-emissions trade-offs for a particular configuration.

6.3 Ceramics have been shown to be capable of operation as sliding or bearing surfaces in valve trains, with a reduced wear rate and a very significant reduction in the amount of lubrication required. Furthermore, the consequent reduction of mass should allow the valve train duty to be uprated.

6.4 The operation of ceramic bearings in large engine turbochargers should provide improved protection against abrasive and corrosive wear.

6.5 Manufacture of complex shapes, including turbocharger rotors, has been shown to be feasible by injection moulding.

6.6 A zirconia powder has been developed that can be manufactured in the U.K. by electro-refining instead of the chemical co-precipitation technique used by overseas suppliers. Control of the chemistry and particle size has produced a powder eminently suitable for sprayed thermal barriers.

As Hondros (12) and others have pointed out, the widespread euphoria of a few years ago concerning the future ceramics market has been replaced by a sober realism in which the prime requirements are not only to demonstrate a net benefit, but to increase reliability and reduce cost. I believe that the above achievements by the CARE programme will make a major contribution to the satisfaction of this real market. Clear advantages of using ceramics have been demonstrated and near net shape methods of manufacture will do much to reduce the cost towards a level acceptable to the engine manufacturing industry.

7. REFERENCES

1) GODFREY, D.J. and MAY, E.R.W. 'The Resistance of Silicon Nitride Ceramics to Thermal Shock and Other Hostile Environments' in Ceramics in Severe Environments, Ed. W.W. Kriegel and Hayne Palmour III, Plenum Press, New York, 1971, p.149.

2) GIBSON, B.D. and STONE, R.B. 'The Design and Performance of a Ceramic Piston', Proc. Conference on Non-Metallic Materials for the Royal Navy, Plymouth, 1975, Paper 2.

3) PARKER, D.A. and SMART, R.F. 'An Evaluation of Silicon Nitride Diesel Pistons', Proc. Brit. Ceram. Soc., 1978, No. 26, p.167.

4) KAMO, R. and BRYZIK, W. 'Adiabatic Turbocompound Engine Performance Prediction', SAE Paper 780068, 1978.

5) BRYZIK, W. and KAMO, R. 'TACOM/Cummins Adiabatic Engine Program', SAE Paper 830314, 1983.

6) A Programme for the Wider Application of New and Improved Materials and Processes (NIMP), 1985. The Report of the Collyear Committee (HMSO).

7) WARD, D. 'Hot Heads Take Over in a Diesel Engine Revolution', The Engineer, July 1982, p.22.

8) HOLMES, K. 'Increasing Diesel Efficiency by Changing the Heat Balance', Automotive Technology International, 1987.

9) MIYAIRI, Y. et al. 'Selective Heat Insulation of Combustion Chamber Walls for a DI Diesel Engine with Monolithic Ceramics', SAE Paper 890141, 1989.

10) MOREL, T. et al. 'Heat Transfer in a Cooled and an Insulated Diesel Engine', SAE Paper 890572, 1989.

11) BLACKBURN, S., HEPWORTH, M.A., KERRIDGE, C.R. and SENKENN, P.G. 'Toughened Zirconia Ceramics from Electro-Refined Powders', United Ceramics Ltd., Stafford, 1987.

12) HONDROS, E.D.   'Opening Address to the Colloqium on Designing Interfaces for Technological Applications: Ceramic-Ceramic and Ceramic-Metal Joining', Proceedings edited by S.D. Peteves, Elsevier, 1989, p.xi.

Overview of the Advanced Ceramics for Turbines (ACT) programme

A.Bennett

Rolls - Royce plc., Advanced Technology Centre, University of Warwick, Coventry CV4 7AL

ABSTRACT: A brief review of the background to the formation of the Advanced Ceramics for Turbines programme is presented, together with its objectives and the programme highlights.

1. PROGRAMME INITIATION

During the 1950's and 1960's British research efforts, supported by Government funding, were at the forefront of world developments in the field of advanced ceramic materials development and application. The energy crisis years of the 1970's, and their effects on the UK economy, caused the industrial partners to withdraw from this work, and and from many other areas of long term research. The world lead in ceramics was taken over by America, where pioneering work was supported by the Department of Defence and the Department of Energy. During the early 1980's, this lead has been shared with Japan, where government support has been made available through MITI, the Japanese Ministry for Information, Trade and Industry. These nationally promoted programmes in America and Japan have concentrated heavily on developing fuel efficient reciprocating and gas turbine engines, primarily for automotive use in the civil and military markets.

Prompted by continuing interest in Rolls-Royce and other potential end user companies, and the rapid pace of developments in America, Japan and also in some European countries, a national discussion on the means to revitalise the UK advanced ceramics industry was initiated in 1982. The main candidates for exploitation were seen to be the reciprocating and gas turbine engines, as recognised in the rest of the world, but it was also noted that the design problems, operating environments, and consequently, the materials of most interest in these two applications were significantly different. Two groupings were thus formed, the Advanced Ceramics for Turbines group, and the Ceramic Applications in Reciprocating Engines group, each able to concentrate on its own problems, but sufficiently in contact with one another to be able to exchange information on areas of common interest.

It was clear from the general state of the UK economy, and the current profitability of potential UK industrial suppliers of advanced ceramics, that to induce investment in long term, high cost, high risk research work, considerable pump priming finance would have to be made available. The Department of Trade and Industry made funds available to assist each group to establish its objectives, to exchange information and ideas, to

propose a research programme, and subsequently to coordinate and monitor
individual projects.

The discussions within the ACT group continued between three
sub-groupings; the end users, the material suppliers, and the national and
Government research establishments. A suite of twelve research projects
were submitted to the DTI, and were accepted for funding assistance up to
half the project costs. The first project began in 1984, at which time
the overall programme had a total budget of £2.2m.

2. PROGRAMME OBJECTIVES

The main objectives in the design of gas turbine engines continue to be to
reduce the cost of ownership, and to increase performance. These
objectives manifest themselves in terms of reduced fuel consumption,
longer periods between overhaul, and increased thrust to weight and thrust
to size ratios. Among the requirements these objectives impose are
improvements and retention of high strength for long times at high
temperatures, increased stiffness and reduced density. The major benefit
of using ceramics is the potential to increase overall engine efficiency.
If compressor delivery air is diverted from the core of the engine, and
used to cool metallic structures at the back of the engine, it is unable
to take part fully in the engine cycle and generates little thrust. Any
reduction in this requirement, made possible by the use of more refractory
materials is highly desirable. Additional benefits of using ceramics
include the reduction of stresses at high rotational speeds through
reduced material density and therefore lower component weight, and a
reduction in design complexity and total engine component parts count, by
using solid ceramic components instead of sophisticated internally cooled
metal structures. These effects of using ceramics cascade down into
second and third order effects including the use of lighter and less
stressed airframes when carrying lighter engines, and less fuel. The
total potential benefit of ceramic materials is thus considerable.

The major barriers to using ceramics are no less considerable, although
they are well known and thoroughly documented. First there is the
inadequacy of current materials to provide the required mean behavioural
capability. On the chemistry front there is long term creep and
oxidation, and on the manufacturing front, the problem of generating the
same high strengths in complex components as can be had in small test
pieces. The second problem is one of inconsistency in behaviour from one
sample to the next. Ceramics need quality control procedures at least an
order of magnitude more stringent than used in the best metals practice,
and these are simply not available. The seemingly inevitable existence of
random defects in the materials as manufactured, and an extreme propensity
for picking up damage in use, mean that their behaviour, both as
manufactured, and with time, are not readily predictable. The third major
problem is a lack of design methods which take account of the
idiosyncrasies and problems of brittle materials, and allow their
strengths to be exploited with confidence. Without adequate and
predictable properties, such methods, leading to optimised, material and
process efficient designs cannot be developed properly, and without these
tools, the materials can not be made to function effectively.

The objectives of the ACT programme were thus clearly set in terms of
providing the material chemistry, and the manufacturing technology,
including the quality control, to produce consistent, high performance

materials in component form. The generation of a design philosophy was not seen as appropriate within the context of the initial ACT programme, but would be a vital area for future work.

The end user group included Noel Penny Turbines, who manufacture relatively small, short life time engines, Rolls-Royce, who manufacture engines with life requirements of a few thousands of hours, and G.E.C., who expect several tens of thousands of hours of service from their engines. Reliability, operating temperatures and load bearing requirements also vary considerably with application within each of these environments, which compounds the difficulty of setting simple, specific, property targets. Because of the diversity of interests and applications within this group, each end user set their own objectives, and presented them to the rest of the membership. From these consultations, each member was able to deduce and set the targets for their own activities, and then address them, through a project proposal. Although the extremely severe turbine blade target properties were adopted by several members of the group, it was recognised that any material behaviour or processing improvements, particularly if they related to reliability, would bring potential applications and benefits to all the end users.

3. PROGRAMME CONTENT

Having agreed in principle to support a programme, the DTI required that all potential projects be submitted for evaluation by a selection committee, which would include representatives from industry and from various Government research establishments. The outline proposals were discussed in open meetings of the whole ACT group, and were then refined by their sponsors, and submitted for consideration by the DTI committee. From the committee review process emerged an integrated suite of projects, covering a wide range of materials and manufacturing problems. The projects were submitted to further scrutiny by the DTI using standard criteria, before public funds were approved.

The programme has supported 15 projects in total, which have covered a wide field of activity. These included a project on silicon nitride powder manufacture led by Harwell, two projects on silicon nitride composition development led by Turner and Newall (T&N) and by Associated Engineering Developments, (AED, since bought by T&N),and three projects on component manufacture in silicon nitride, led by Fairey Tecramics with British Ceramic Research Association (BCRA, now BCRL), by Lucas Cookson Syalon (LCS, now Vesuvious Zyalon Midland), and by AED.

There have been six projects looking at other materials, two of which were concerned with fibre reinforced glasses and glass-ceramics, led by Harwell and Pilkington Brothers, with two projects on ceramic matrix composites and one on sintered silicon carbide, all led by T&N. The twelfth project was on non destructive evaluation (NDE) techniques, led by Harwell, and these comprised the early core of the programme.

When the first two reinforced glass and glass ceramics projects ended, they were replaced by second phase activities, again under the leadership of Harwell and Pilkington Brothers. The final project is on plasma sprayed zirconia coatings, and is being undertaken by a consortium led by General Electric Company (GEC).

Each project has been reported formally through the DTI to a monitoring panel, to ensure that scientific value for money has been maintained, and that the projects are being properly directed and conducted. In addition, each project leader has regularly had to present his work to the rest of the ACT group, and to representatives from several Government research establishments not directly involved in the ACT programme.

4. PROGRAMME HIGHLIGHTS

To report in detail on the progress of all 15 projects is clearly beyond the scope of this paper, but it is possible to look briefly at the main findings.

4.1 Harwell Powder Project

This project aimed to use cheap non toxic starting materials to produce carbon doped silica sols, and to convert them by carbothermal reduction in nitrogen, into alpha silicon nitride powder. This is essentially similar to the commercial process used by Toshiba, but with subtle variations to improve powder purity and compressibility. The project produced powder of a good quality, but only in laboratory scale quantities, making a full evaluation of its potential impossible. The project yielded much valuable information on powder behaviour and selection criteria, as well as generating manufacturing data. As yet the process has not been scaled up to produce useful quantities of powders.

4.2 T&N Silicon Nitride Project

This project looked at novel sintering aids, and in particular at neodymia. The creep capability achieved was comparable with current materials, but still an order of magnitude below the required level. Bend strengths at 1350 C were in the range 350-450 MPa with Weibull "m" of 10. Oxidation at 1400 C was far better than for most high temperature silicon nitrides, many of which contain alumina and yttria, and experience severe oxidation at temperatures above 1300 C.

4.3 AED Silicon Nitride Project

This project looked at novel sintering additions, based on the results of earlier work done by BCRA as a multi-client project. One composition evolved contains alumina, yttria and other additives, and is suitable for pressureless sintering. This material yielded bend strengths of 800MPa with "m" of 10 at room temperature, but less than 300 MPa and suspect life at a temperature of 1400 C. A second material based essentially on a low yttria addition and densified by hot isostatic pressing (HIPing), yielded bend strengths in line with the target specifications at all temperatures, although time dependant properties were not evaluated.

4.4 Fairey/BCRA Manufacturing Project

This project looked at the fundamentals of the injection moulding process, and produced the following major conclusions. The binder system used should have as wide a decomposition range as possible, and the volume fraction of ceramic powder in the moulding mix should be as high as possible. Within these two constraints, the mix should have a low viscosity at the injection temperature, as many of the typical defects formed result from high stresses induced during the moulding process. To further minimise defects and binder removal problems, the binders and powders used, and the mix homogeneity must be controlled to give a high level of uniformity from batch to batch, and powder surface area is of particular importance in this respect. Finally, die design is critical to success, and in particular, the use of as large an in-gate size as practicable.

4.5 LCS Manufacturing Project

This project studied the dry bag isopressing of large diameter ball bearings, and the injection moulding of turbine blades. Each component was manufactured in a LCS proprietary material. The isopressing work involved extensive powder processing developments which led to a free flowing "soft" granular material, which facilitated a successful manufacturing process. The injection moulding study concentrated on the development and characterisation of a binder system, and a removal process. Extensive NDE of trial mouldings detected no evidence of defects. Moulded test bars yielded bend strengths of 750MPa, with "m"= 8. The moulded aerofoils were close to drawing, showing more work would be required to give accurate nett shapes, but the deviations from nominal form were small and consistent.

4.6 AED Manufacturing Project

This project examined at several techniques for cold forming complex shapes. Slip casting was found to be extremely difficult to control adequately,and limited success was achieved. Freeze drying processes showed much promise, although the drying procedure is long and costly. Some progress was made on cold isopressing complex shapes directly, but the greatest success was achieved from machining shapes from cold isopressed (green) billet.

Attention to detail is essential in the mixing and isopressing before machining, and in the densification routine, to obtain the best results, and high precision tooling is needed. The shape forming step however is essentially one of CNC programming so modifications and completely new designs can be introduced relatively easily and at low cost, making for excellent flexibility of the process. With pressureless sintering, less than 0.7% dimensional variability is possible. With HIPing, less than 0.3% variability was achieved on a complex aerofoil, which was within the required tolerance, providing a true, nett shape capability.

4.7 Harwell Glass-Ceramic Project

This project looked at the manufacture of borosilicate glass, uniaxially reinforced with Nicalon silicon carbide. Using a hot pressing technique, 40% by volume of fibre could be incorporated, and flexural strengths of over 1000MPa with "m" of 20 were measured. The material showed high work of fracture, extensive fibre pull out, giving rise to a degree of pseudo ductility.

4.8 Pilkington Glass-Ceramic Project

This project, linked to the Harwell project, aimed to select a narrow range of glass-ceramic materials to use as higher temperature matrices in a fibre reinforced system. This work identified a variety of matrices,

and some of the key processing parameters required to produce viable
materials. The combination of the behavioural data from Harwell, and the
materials development at Pilkington's, prompted both groups to submit
proposals for a second phase of activity. These proposals were accepted
and will be described later.

4.9 T&N High Performance Composite Project

This project focused on silicon carbide whisker toughened silicon nitride
materials. Several matrix compositions were used, with different types
and levels of whisker additions and densification processes. Marginal
improvements were obtained in strength, "m", fracture toughness and high
temperature strength retention, but batch to batch variations due to
processing difficulties could not be overcome. The toxicity and the cost
of the whiskers, together with the processing problems, were felt to
present an overwhelming argument against these materials given the modest
benefits achieved.

4.10 T&N High Temperature Composite Project

The prime objective of this project was to study the effects of
introducing chopped fibres into ceramic matrices. The initial work looked
at chopped Nicalon silicon carbide. This proved extremely prone to the
formation of intractable clumps which defied all attempts to redisperse
them. The project was cancelled prematurely because of the total
inability to devise a sensible processing route for these materials.

4.11 T&N Sintered Silicon Carbide Project

Dense sintered silicon carbide was produced using the boron plus carbon
(GE patent) sintering system. All available grades of beta silicon
carbide and several sintering aids were studied, and a number of promising
materials were produced. All exhibited excellent resistance to oxidation,
and retention of strength at 1400 C.

As a result of the success of this project, T&N have licenced the GE
process, and have a fully commercialised material on offer, which is
comparable in terms of strength and oxidation behaviour, with other
materials available from sources outside the United Kingdom.

4.12 Harwell NDE Project

One of the most frustrating aspects of NDE of ceramics is not so much the
inability to detect defects, but the inability to interpret inspection
data in terms of likely material behaviour. This project made significant
progress in refining and developing ultrasonic, thermographic, and x-ray
techniques. Pores, cracks, high and low density inclusions, and density
banding defects in samples from several sources have been observed, at or
below detection limits previously reported, and inspection routines for
many typical shapes can be fully specified. Correlation of NDE findings
with actual behaviour of components remains less well defined.

4.13 Harwell Glass-Ceramic Project (Part II)

This project is still ongoing, and is looking at several aspects of
manufacture and behaviour, including sol-gel methods to prepare the

matrices, and coatings for the fibres. Work is also being done to support the optimisation of pressing and heat treatment routines being pursued in the Pilkington project.

4.14 Pilkington Glass-Ceramic Project (Part II)

This project, to develop high temperature composites has yielded materials with 500MPa useful strength at 1200 C (tested in air), by the hot pressing and heat treatment of reinforced glass-ceramic matrices. Control of the process, matrix cracking and oxidation of the Nicalon silicon carbide fibres remain as problems to be solved, but the materials produced show considerable promise.

4.15 GEC Plasma Sprayed Zirconia Project

This is a collaborative project currently in progress, involving Universal Abrasives, Plasma-Technik, BCRL, Professor W. Steen formerly of Imperial College, now of Liverpool University, Rolls-Royce plc., and led by GEC. Its objectives are to develop improved plasma sprayed zirconia thermal barrier coatings, with resistance thermal cycling and aggressive gas turbine environments. Although little detail can be made available at this stage, initial results look promising.

5. PROGRAMME REVIEW

The programme has yielded much that is positive, such as the development of a UK source of quality silicon carbide material, some valuable silicon nitride materials, improved and flexible nett shape manufacturing processes, and some new NDE capabilities. There has been a general increase in the level of understanding of the materials processing and behaviour, which will be of significant importance in the future, and the ongoing projects look set to produce an extremely interesting reinforced glass-ceramic material and a much improved ceramic coating system.

Inevitably, some of the specific project objectives were found to be over optimistic in relation to time scale, but never the less, all the projects yielded new information, and in many cases, positive results. The work has not been easily accomplished, and there have been differences of opinion within the group, both over specific details and over broad concepts. This has produced an exciting and stimulating environment, charged with a gentlemanly adversarial spirit in which progress has been rapid, and discoveries have been frequent and valuable.

All the participants in the programme have made dramatic progress in their understanding and capabilities, and this general enlightenment, though difficult to quantify, must be recognised as one of the significant benefits of the programme to the country as a whole, and a major validation for government involvement in research of this kind.

The ACT programme steering committee has now been disbanded, but the Government's new LINK initiatives in structural composites and high temperature, high performance materials offer the possibility for a wider range of participants to enter into collaborative research projects in this field. The evidence to date suggests the individual LINK programmes will include many of the features of ACT, and it is to be hoped that they too will produce such a wealth of enlightenment and of exploitable technology.

6. ACKNOWLEDGEMENTS

The author would like to thank Rolls-Royce plc for granting their permission to write and present this paper, all the individuals who have been involved in the ACT programme for their kind assistance and comments on the original drafts, and their parent companies and the Department of Trade and Industry, for their agreements to release the information contained.

Paper presented at Conf. on Ceramics in Energy Applications, Sheffield, April 1990
Session 2B

Application of ceramics for hot gas radial fans for temperatures up to 1250°C

H.-J. Barth, M. Gür, S. Morgenroth, R. Scholz

Institut für Maschinenwesen, Institut für Energieverfahrenstechnik, Technische Universität Clausthal,
3392 Clausthal-Zellerfeld,
Rep. of Germany

ABSTRACT:
Design work and experiments are discribed which were realized with the aim of developing fans for the transport of hot gases with temperatures up to 1250 °C. Ceramic-Steel compound fan was designed, produced and tested. It consists of an air cooled steel disk bearing the centrifugal forces and single ceramic blades, serving as elements of gas transport and heat protection. In this compound fan the element "blade" shall undergo tests which mean an intermediate stage on the way to a fan which is produced completely from ceramic material.

1. INRODUCTION

For many hot gas processes it is desirable to circulate mechanically hot gases with temperatures up to 1250 °C in order to improve the process control or equalize temperature gradients in reactors. Actually only steel fans are available with a sufficient creep strength only for gas temperatures up to 800 °C. Therefore our problem was to find design solutions permitting a mechanical gas circulation also at 1250 °C. Soon it was found out, that a satisfying temperature resistance could be achieved only by the use of ceramic materials. Discussions with experts from universities and industrial companies as well as a literature survey made clear that despite the often euphoric expectations in modern ceramic materials there exist a series of problems restraining their application. One of these is the substantial lack of experience in mechanical engineering with the design, production and strength qualities of ceramic parts, on the other hand many producers and operators of hot gas plants showed a vivid interest in devices for a hot gas circulation at high temperature levels which encouraged us to take up and continue the investigations presented here. We thank the German Company For Resarch (DFG) for their support of this work.

2. CERAMIC - STEEL COMPOUND FAN

2.1 DESIGN PROBLEMS WITH CERAMICS

High performance ceramic materials demand in the same way as

other materials special design rules adapted to their material properties and the intended manufacturing process. Only a limited number of tested rules exist. Special difficulties result from the fact that ceramic materials are not ductile. Local stress peaks which practically can not be avoided in those areas where forces are introduced into a part must be equalized by the aid of plastic intermediate layers. While compression stresses (except in those areas where the load is introduced) are born without problems, tensile stresses should be kept as low as possible. This is valid for mechanical as well as for thermal stresses. Fundamentally for ceramic parts only a statistical value for the probability of survival can be determined. The quality of produced parts depends largely on :
  - the purity and distribution of grain sizes of the raw
    powder
  - the design shape (wall thickness, notches)
  - the handicraft experience and skill in the production
    process.
The achieved accuracy of size depends on the used material. In most cases a considerable shrinkage is observed. Simple machining by milling, turning or drilling is possible only before sintering with the danger of producing micro-cracks. After the sintering process only grinding can be used for any change of shape or size, and expensive grinding tools are necessary because of the extreme hardness of ceramic parts. When the size of parts is increased the danger of defects grows as well. For that reason producers recommend – again depending on the material and its sintering process – maximum values for the over all dimensions of parts. Structures of greater size therefore can be produced only from separate parts mounted together in an appropriate way. Advantage of high performance ceramic materials are above all their excellent temperature resistance and their resistance against abrasion and corrosion. For applications in rotating systems their lower density in comparison with that of metallic materials is favourable, because lower centrifugal forces are produced and consequently lower centrifugal stresses must be born. There are great differences between some of the material proporties of ceramic materials, thus influencing characteristics of parts produced from them. So a low thermal conductivity permits to use ceramic parts as a heat protection for a cooled metal structures. However at the same time strong temperature gradients are induced causing thermal stresses which might lead to premature failure. For applications under high temperatures the evaluation of materials is rendered difficult by lacking values as their measurement at high temperatures is complicated and has not yet been realized for all available materials.

2.2 DESIGN OF CERAMIC - STEEL COMPOUND

After a systematic investigation of possible devices for the circulation of hot gases radial fans were found as an adequate solution. A first design was chosen which allows to test

single ceramic blades. It consists of a heat resistant metal
disk cooled by compressed air (fig.1). The air flows through

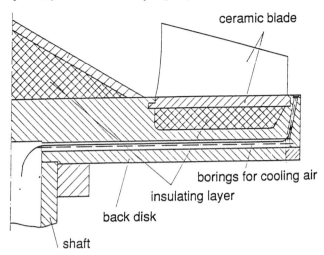

Fig.1: Ceramic-steel compund fan, shematic arrangement

the hollow shaft to the disk, through radial borings to the
circumference and leaves the wheels through numerous small
borings on the front side. Considering the energy situation
this mixing of cooling air with the hot gas is undesired as
gases as hot as possible should be circulated. The mixing was
tolerated here because the production of the wheel became much
easier than with a closed cooling system. The ceramic blades
under investigation are inserted into the metal disk. The disk
thus has to take over the centrifugal forces acting on the
blades on the outer border. The inner ring slot prevents
slipping out of the blades. Between blades and wheel a ring
shaped hollow space is left which is filled with a heat
insulating layer of Al_2O_3 - fibers. A conical hood insulated
as well improves the flow conditions near the gas inlet. For
the set-up in fig.1 temperature fields were calculated by
means of a FEM- program. It was found out that for the metal
disk the highest temperatures had to be expected on the outer
border, that means in that area where the highest mechanical
stresses would act.
The idea of compound design of steel disk bearing the main
part of stresses and set of separate ceramic blades insertet
into the disk also takes into*the principle of a design with
neutral behaviour in case of damage. If one or the other of
the blades suffers small damages, the complete set of blades
will remain fixed in the slots of the steel disk. Complete
failure will occur only after severe damages.

2.3 DEVELOPMENT OF CERAMIC BLADES

Table 1 gives a survey on the important data of the
*consideration

investigated blades. Type A mentioned there discribes a
commercial metal-wheel produced in quantity for hot gas
applications up to 800 °C. Its delivery characterizes the state
of art and serves as standard of comparison for the blades to
be developed (see ch. 3.1).

Table 1: Characteristic sizes of investigated impellers

impeller Type	frontplate	d_2/d_1 [-]	d_2 [mm]	β_1 [°]	β_2 [°]	z [-]	blade material
A	yes	-	630	-	90	10	steel
B	no	2	550	34	54	18	RBSN
C	no	2	550	90	90	11	SiSiC

d_1 blade inlet diameter
d_2 blade tip diameter
β_1 blade inlet angle
β_2 blade outlet angle
z   number of blade

At the beginning of the design of a ceramic blade its shape
must be fixed. It is determined by the
- flow geometry and the rsulting delivery
- the acting forces
- the available technology of production
- the demands of simple assembly

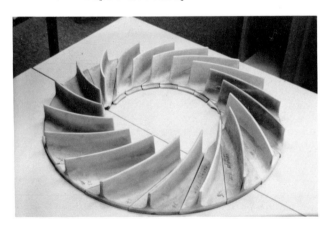

Fig.2: Set of RBSN-blades with backward bend before assembly

Engines and Fans

As separate blades were to be inserted into slots of the steel
disk, gliding between disk and blades as well as between the
blades themselves must be possible.

After a thorough discussion with experts as material for the
first blades RBSN (reaction bound silicium nitride) was
chosen, having an especially low shrinkage. For that reason
it was expected that after sintering no machining would be
necessary. Together with the producer we agreed on a blade
geometry (Type B in tab.1 and fig.2) with backward bend and
without front disk. In the beginning we were not sure, if
without a front disk a satisfying pressure increase could be

Fig.3: Radial-blades from SiSiC
      above: complete set before assembly
      below: being assembled

obtained. Thus the backward bend was not chosen to reduce the power of the driving motor, because in high temperature plants this power in most cases can be neglected in comparison with the thermal power of the process. The backward bend was intended to assure a minimum delivery. Each blade is situated on a ground plate carrying profiles on the inner* edge which can be insertet into the slots of the metal wheel. The lateral edges of the ground plates are straight to improve gliding movements between the blades. They are considerably inclined versus the radial direction because of the backward bend of the blade. Therefore friction forces act between the lateral edges which can disturb gliding. In addition owing to the bend of the blades the centrifugal forces do not act purely in radial direction. Instead bending is superimposed. Gas forces are so small that they may be neglected. For the evaluation of stresses FEM- calculations were carried out. After some variations of the curves between ground plate and blade the maximum value of tensile stresses could be brought down to ca. 45 N/mm^2 compared with the stress of fracture of 200 N/mm^2 (producers information) for a nominal speed of 3500 r.p.m.

In order to verify this result a photoelastic model of the blade was investigated which had been loaded in a heated centrifuge using the frozen-stress-method. Similar values were found as in the FEM-calculations. Finally strain gauges were applied to two original blades and loaded on the centrifugal set-up, where they broke apart at a speed of some 2300 r.p.m. From the stress values registered continuously up to that speed stresses could be extrapolated corresponding sufficiently to the calculated FEM-values. Further tests also resulted in total failures at lower speeds than anticipated, so that the operating speed for later hot gas trials was set down. Apparently all cracks were initiated close to the inner edge of the blades, where in the FEM-calculations no increased stress values had been found. In this part of the blade a sharp-edged notch is located, more over the stress state in this part obviously is multi-axial. It is well known that multi-axial tensile stresses further the danger of brittle fracture. It is unclear in which way a multi-axial state of stress influences the bearable stresses of a ceramic material. Introducing compression forces into a ceramic part is another problem. While in metallic parts stress peaks produced by local uneveness in the area where forces are introduced, are reduced by plastic deformation, a comparable stress reduction in ceramic parts is not possible. For low-temperature applications plastic intermediate layers are recommended for the zone, where forces are introduced. In the centrifugal trials mentioned above adhesive tapes were used for that purpose. As cracks occured nevertheless, their reason at this moment remains unclear.

Bending stresses in the blades can be prevented by the use of purely radial blades (Type C in tab 1 and fig 3). SiSiC was chosen as material. The number of blades was reduced from 18 to 11. With the synchronous choice of a new material and a new

*and outer

geometry the rule was violated which says that in one step
only one parameter may be variied. In this case a new material
was tolerated, because in this way a new set of blades became
available in a short time. However we plan to investigate as
well another set of blades of the Type C made of RBSN to
complete the program of investigation. Several blades of the
type C from SiSiC were successfully tested at speeds up to
2500 r.p.m. and could be inserted into the metal wheel
discribed in chapter 2.2, as preparation fortrials in a test
plant for hot gas fans.

3. TEST RESULTS:

3.1 TESTS WITH A STEEL-HOT GAS FAN

A fan test plant was established and was integrated into a
combustion plant (Fig.4) with a thermal power of 0,5 MW. The
test ducts and the volute casing are lined with refractory
material. The generation and the circulation of waste gas with

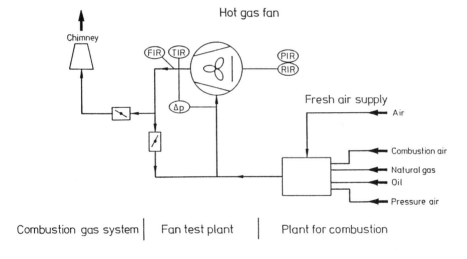

Fig.4: Experimental plant

a flow rate of up to 5000 m³/h (for instance from natural gas
combustion) and temperatures till 1400 °C mean, that the
results of tests are transferable to later industrial
applications. The hot gas can be circulated in order to reduce
energy costs. In order to test the plant and it's measuring
devices a commercial welded hot gas fan was first of all
investigated. This fan was made from a heat resistant
austenitic steel (type A in table 1) and has a drive with
stepless adjustable rate of rotation. The test of the fan

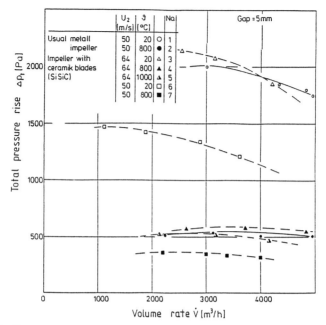

Fig.5: Comprasion of characteristics of the metal impeller
        Type A and the compound impeller Type C

yields the fan delivery shown in Fig. 5. This was considered
as an aim for the delivery of ceramic blades to be developed.
It was tried to obtain criteria for a damage recognition by
continuous registration of vibrations on one of the bearing
housings of   the fan.   Thereupon   three   identical   metal
impellers were run at higher than the calculating fan speed
and - temperature until failure. All three impellers failed
after a duration of test close to the predicted lifetime due
to crarks at the junction of hub to blade. The cracks lead to
a   considerable   increase   in   the   amplitudes   of   vibration
(Fig.6), while the delivery remains unchanged.

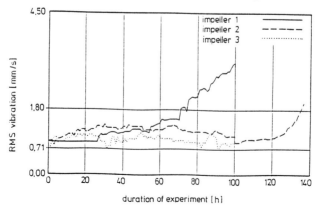

Fig.6: Vibration behaviour of the metal impeller Type A
        during long time running tests

Though for a ceramic-steel compound fan different reasons of failure must be expected, the measuremet of vibrations was planned as well for the following investigations of compound impellers.

3.2   INFLUENCE OF THE GAS FLOW ON THE SHAPE OF COMPOUND IMPELLERS

Ceramic materials are applied with the intention to cross over now existing temperature limits. Therefore rules for the design of blades for the special flow conditions of hot gases do not exist. As the gas density at high temperatures is low, only a limited pressure increase can be expected.On the other hand the necessary drive power remains low, so that a high efficiency becomes less important. The toughness of hot gases is considerably higher than at lower temperatures. The boundary layer of gas flows therefore must be greater. A reduction of leakage can be expected. As the size of ceramic blades is limited, a sufficient pressure increase despite a lower impeller size must be obtained by a higher peripheral speed. In order to gain experience on the design suited for flows in hot gas applications, questions as the suitable choice of the blade inlet- and outlet angle, delivery characteristics without sideplate and the influence of the impeller main size and gap geometry were to be investigated in tests with the compound fan. Compromises in the shape of

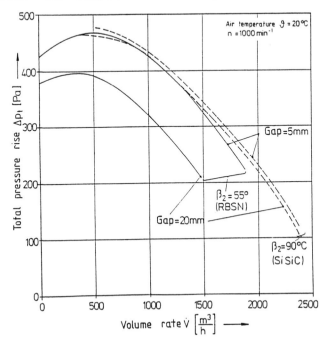

Fig.7: Characteristics of the compound impeller Type B and C

blades because of problems of production or strain must as well be compensated by higher peripheral speeds.

3.3   TEST RESULTS COMPOUND FAN IMPELLERS IN HOTGAS

Only at ambient temperatures some measuring points could be recorded with the RBSN-blades. In trials realized in hot gas lateron two blades were destroyed, when the first resonance rate of rotation was reached. One of the blades obviously failed because of a fault of material, the second was presumably destroyed by fragments flying away. The refractory lining of the volute casing remained undamaged. The originally feared effect, namely that a blade breaking off would distroy the whole set of blades, did not occur. The disturbed equilibrium produced by the damage, did not produce consecutive harms. The fan could be taken out of operation without problems. In a next test temperature measurements were carried out on the impeller C in order to determine the necessary flow of cooling air. The impeller stood still in these tests. A flow of 5 % of the total gas flow was found The delivery of the impeller C is shown in Fig. 5 and 7 in comparison with the delivery of the metal impeller A. In Fig. 7 the impeller B and C are compared at 20 °C for different gap widths between impeller and casing. We see, that at the same peripheral speed with the compound impeller C less pressure increase can be achieved with than the metal impeller A. But the difference can be compensated by a moderate increase of the peripheral speed of the compound impeller. With this higher peripheral speed the compound impeller could be run as well at 1000 C. Continuous tests at 1000 °C have not yet been carried out. Fig. 7 shows clearly that in spite of its complicated form the impeller B does not produce a higher pressure increase than the impeller C. This encourages the use of simple shapes for ceramic blades for such impellers.

4. CONCLUSION

Decisive for the succesful use of ceramic parts for hot gas fans is a design adapted to the material and its manufacturing process. So above all an efficient technique of assembling ceramic parts and ceramic-metal compounds must be developed in a greater extent. The compound impellers described above served for the test of ceramic blades. Their temperature limit is determined by the temperatures of those parts of the metal wheel, which are exposed directly to the radiation. Impellers for higher temperatures can only be realized, when all metal parts can be screened perfectly with ceramic parts. Metal parts always must be cooled, which causes thermal stresses. Therefore they do not represent an ideal solution. The aim of further developments is therefore a pure uncooled ceramic construction. Evidently a series of development steps will be necessary before hot gas fans for temperatures up to 1250 °C with sufficient reliability will be available for different industrial high temperature processes.

Comparison of techniques for the measurement of the emittance of ceramic materials

J D Jackson, E Romero and J J Norris

Engineering Department, University of Manchester, Oxford Road, Manchester M13 9PL.

ABSTRACT: The paper reports radiometric measurements made using three different techniques of the total emissivity of a number of specimens of a refractory material (composition 62.3% Al_2O_3, 32.1% SiO_2) covering the temperature range from about 150°C to 1000°C. The main aim of the work was to compare different measurement techniques with a view to seeing whether consistent results could be obtained.

Two of the methods used in the present study were of the 'direct' type in which the intensity of radiation received by a calibrated total radiation pyrometer from a heated specimen was used to determine the directional total emissivity. The two 'direct' methods differed only in that in one case the specimen emitted thermal radiation to the surroundings under steady state conditions whereas in the second case the process was a transient one. The third method was of the 'indirect' type in which an unheated specimen was subjected to a black body radiation field of known intensity by traversing it through a small hole into a large heated cavity. A total radiation pyrometer, which viewed the specimen through the hole, received both the radiation emitted by the specimen and the black body radiation reflected from it.

In the case of each of the three techniques it was possible to vary the angle of view of the radiation pyrometer with respect to the normal to the surface of the specimen. Directional measurements made at various fixed temperature levels showed little variation with angle of view, which indicated that the material was a diffuse emitter.

The emissivity values obtained using the three techniques were found to be in good agreement with each other. Excellent consistency was found in the case of measurements made using different specimens. A well-defined distribution of emissivity as a function of temperature has been established for this material. The emissivity is about 0.9 at 150°C and decreases with increase of temperature to about 0.55 at 1000°C. This behaviour is consistent with the material having a very non-uniform distribution of spectral emissivity of the kind which has been reported earlier for alumina and silica based materials.

NOMENCLATURE

$e_{b\lambda}(T)$ Spectral value of the hemispherical emissive power from a black body at absolute temperature T.

$i(\theta)$ Intensity of total thermal radiation in the direction θ.

T_a Absolute temperature of the 'ambient' temperature enclosure in the arrangement for the 'direct' measurements of emissivity.

T_b	Absolute temperature of the heated black body cavity in the arrangement for the 'indirect' measurements of emissivity.
T_s	Absolute temperature of the target area of the specimen surface.
λ	Wavelength of the radiation.
σ	Stefan Boltzman constant.
θ	Angular coordinate in degrees measured from normal to the specimen surface.
$\alpha_\lambda(\theta,T)$	Directional spectral value of the absorbtivity of a surface at temperature T to thermal radiation incident in direction θ.
$\epsilon_\lambda(\theta,T)$	Directional spectral value of the emissivity of a surface at temperature T in the direction θ.
$\bar{\epsilon}(\theta,T)$	Directional total value of the emissivity of a surface at temperature T in the direction θ.
$\rho_\lambda(\theta,T)$	Directional spectral value of the reflectivity of a surface at temperature T in the direction θ.

1.   INTRODUCTION

Ceramic materials are widely used in high temperature applications such as furnaces, combustion chambers, heat shields, etc. where thermal radiation is a dominant mechanism of heat transfer. Thus there is a real need for data concerning the radiative properties of such materials. In the case of materials operating at very high temperature, questions arise as to the extent to which they are transparent to thermal radiation (Hobbs and Folweiler (1966)). This matter is of particular concern in the case of certain modern materials such as ceramic fibre furnace linings, or high porosity expanded ceramics, which are now finding increasing application (see for instance Fiveland (1985)).

The determination of the emissivity of ceramic materials is not a straightforward matter. Severe practical difficulties arise in connection with the measurement of test specimen surface temperature using thermocouples embedded just below the surface. These thermocouples must be extremely small, and capable of withstanding extremely high temperature. As a consequence of the fact that ceramic materials are, relatively poor conductors of heat, steep temperature gradients exist within a specimen normal to its surface. Thus, the thermocouples should be positioned as near to the specimen surface as possible, and their locations must be precisely known. Furthermore, they should be installed so as to lie, as far as possible, along isotherms in order to minimise errors due to the effects of conduction along the wires and sheath.

It is extremely difficult to achieve uniformity of temperature over the surface of a ceramic specimen operating at high temperature and it is probably for this reason that the calorimetric method of emissivity measurement has not found favour in the case of such materials and attention has instead been concentrated on radiometric techniques. However, in the case of the radiometric approach, other difficulties present themselves. These stem from the fact that the spectral distribution of emissivity can be very non-uniform in the case of ceramic materials (see for instance Dunkle and Gier (1963)). In the far infra-red, the spectral emissivity is uniformly high (greater than 0.9) but, in the middle part of the thermal range (at a wavelength of about 7 microns) a sudden fall in emissivity occurs to a value of about 0.5 and there are sharp peaks and troughs in the spectral distribution at lower wavelengths. Thus the measurement of total emissivity by radiometric means necessitates the use of infra-red measuring instruments which have a uniformly flat response to thermal radiation over the entire thermal range. The use of an instrument having an insufficiently wide acceptance waveband can lead to big errors in measured emissivity.

The alternative approach of measuring the spectral distribution of emissivity and integrating it over the entire thermal range to obtain the total value of emissivity necessitates the use of an optical system with fine resolution and sophisticated electronic equipment to amplify the correspondingly low level output.

Thanks to modern developments in instrumentation and electronics, the equipment needed for making reliable measurements of either total or spectral emissivity is readily available these days. However, full advantage has so far not been taken of this and the topic of emissivity measurement has been a rather one neglected in recent years.

There is certainly a need for improved data on the emissivity of ceramic materials. Although the early studies date back over thirty years, an examination of the literature reveals that emissivity data are sparse and in some cases the published values are contradictory. An early compilation by Gubareff et. al. (1960) quotes emissivity values ranging from 0.7 at 400°C to 0.4 at 1100°C for a ceramic coating on stainless steel. Wade and Slemp (1962) measured the total emittance of several refractory materials and ceramics and give values for silicon carbide ranging from 0.94 at 350°C to 0.78 at 1000°C. Pears (1962) measured the emittance of similar materials and gave values for silicon carbide between 0.9 and 0.65 over the temperature range 300°C to 1600°C and a mean value of about 0.4 for silicon oxide over the temperature range 600°C to 1400°C. Singham (1962) quotes values of 0.66 for silicon oxide at 1100°C and 0.39 for magnesium oxide at 1400°C.

Haas (1964) measured the spectral reflectance of aluminium oxide coatings. Integration of spectral distributions of emissivity deduced from his measurements enables a total emissivity of about 0.3 to be inferred for a temperature of 1000°C.

Cabannes (1970) measured the spectral distributions of reflectivity of several refractory oxides. These distributions have the common feature that the reflectivity is low for long wavelengths (greater than 8 μm) but is much higher for short wavelengths (less than 3 μm). From the data presented it can be inferred that the total emissivity of silicon oxide varies from about 0.9 at 300°C to about 0.7 at 1000°C and that of zirconium oxide from about 0.8 at 300°C to about 0.5 at 1000°C. For aluminium oxide, values obtained are 0.9 at 300°C and 0.6 at 1000°C.

Touloukian (1970) in his compilation of emissivity data gives information for aluminium oxide and quotes a value of 0.8 at 150°C, values between 0.3 and 0.55 at 1000°C and between 0.3 and 0.45 at 1600°C. For silicon oxide he quotes 0.37 at 750°C. He also presents spectral distributions of emissivity for mixtures of these oxides. Integration of these distributions gave total emissivity values between 0.3 and 0.37 at 1000°C and between 0.2 and 0.28 at 1500°C. For a more recent compilation of the radiative properties of refractory oxide materials the reader is referred to Whitson (1976) which deals specifically with aluminium, magnesium and zirconium oxides.

As mentioned earlier, recent interest has been centred on low density ceramic fibre materials. Fletcher and Williams (1984) measured the emissivity of three samples of such material (approximate composition 45% Al_2O_3 and 55% SiO_2) and obtained total emissivity values between 0.93 and 0.96 at 400°C and between 0.51 and 0.66 at 1100°C.

The general picture which emerges from the above review is that the total emissivity of refractory oxide materials decreases markedly with increase of temperature and it is clear that this is a consequence of the very non-uniform spectral distributions of emissivity. This is obviously an important consideration in design and performance estimation of practical equipment such as furnaces and combustors and it has generated an interest in the use of special coatings to enhance the emissivity of refractory linings. The investigation reported here forms part of a programme of work which was in fact initiated to study the effectiveness of certain proprietry coatings applied to refractory material. In this paper we describe some experiments on uncoated specimens which have been carried out to validate our experimental techniques by comparing three different approaches in the measurement by radiometric means of the total emissivity of a 'fired' clay of composition 63.2% Al_3O_2 and 32.1% SiO_2.

2.   'DIRECT' MEASUREMENTS OF EMISSIVITY

2.1 The Method

The 'direct' method of measuring the emissivity of a material by radiometric means is basically very simple, although care must be exercised in the choice of radiometer and in calibrating it.   Thermal radiation from a small target area of the surface of a heated specimen at a known temperature is collected by an infra-red optical system, passes through a chopper and is then incident upon an infra-red detector.   The electrical output generated by the detector is supplied to an amplification system which produces an output voltage which can be related to the intensity of thermal radiation emanating from the specimen surface.   The system must first be calibrated using a source of black body radiation at known temperatures.

2.2 The Radiation Pyrometer

The instrument used in this study was a precision total radiation pyrometer which had a very uniform response over a wide waveband (0.2 to 40 microns).   It was carefully calibrated using several standard reference sources of black body radiation up to a source temperature of 1000˚C.   The relationship between the pyrometer output voltage and the intensity of radiation received by the instrument was established in terms of simple equations by carefully fitting the tabulated results of the calibration tests using least squares polynomial curve fitting techniques.

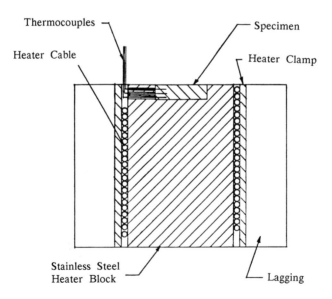

Figure 1.   Details of Test Specimen and Heater for Test Series A
                   Direct Measurements

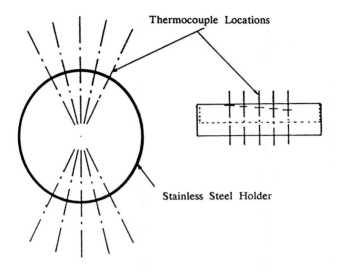

Thermocouple Locations

Stainless Steel Holder

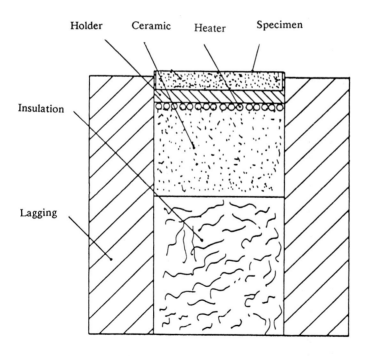

Holder   Ceramic   Heater   Specimen

Insulation

Lagging

Figure 2.   Details of Test Specimen and Heater for Test Series B
Direct Measurements

2.3 The Test Specimens

Two different specimens were designed and manufactured for use in the experiments reported here which were carried in two separate series (Test Series A and B). The specimen for Test Series A was in the form of a small brick 'fired' in a cylindrical cavity of bore 25 mm and depth 6 mm which was machined in the end of a stainless steel bar of diameter 50 mm and length 75 mm (see Figure 1).

Three mineral insulated stainless steel sheathed, chromel alumel thermocouples of diameter 0.5 mm were embedded in the specimen as shown at depths of 2 mm, 4 mm and 6 mm. A coil of mineral insulated stainless steel sheathed cable wound on the outside of the bar and clamped to it by a stainless steel clamp enabled the bar and specimen to be heated by electrical means. The outside was heavily lagged and the whole arrangement was mounted on asbestos supports with the axis of the specimen and heater bar horizontal.

The specimen for Test Series B was in the form of a disc of diameter 50 mm and thickness 5 mm 'fired' in a thin-walled stainless steel cup (see Figure 2).

A mineral insulated stainless steel sheathed electrical heater cable, coiled in the form of a spiral, was kept in good thermal with the underside of the cup by further ceramic material below it. Behind this there was some ceramic fibre and the whole arrangement was encased within rigid, pre-formed ceramic insulation of high porosity and extremely low thermal conductivity.

The arrangement described above was designed for operation at higher temperature than had proved possible in the case of the first specimen with improved measurement of specimen temperature. Ten mineral insulated, stainless steel sheathed chromel alumel thermocouples of diameter 0.5 mm were let in from the side of the cup in pairs at various depths such that the temperature very close to the surface and also at depths of 1.0 mm, 1.5 mm, 2.0 mm and 2.5 mm could be measured. The thermocouple junctions were located on a pitch circle of 5 mm diameter in such a way that they did not interfere with each other (Figure 2).

2.4 Experimental Arrangement and Procedure

The experimental arrangement was as shown in Figure 3. The specimen and pyrometer were contained within a small room which was painted black and had its temperature controlled to achieve a uniform value just above ambient temperature. The specimen was placed at a particular distance from the radiation pyrometer chosen to give a target area in the form of a circle of diameter 5 mm at the centre of the specimen surface when the instrument was focussed on the surface and viewed it normally.

It was possible to vary the angle of view with respect to the normal by adjusting the position of the specimen whilst keeping the working distance between pyrometer and specimen the same. When operating in this mode the target area was, of course, elliptical rather than circular and this placed a limit of 70° on the angle of view to ensure that the target area remained on the central part of the specimen where the temperature was unform and known.

The pyrometer was mounted on a swivel arrangement so that it could be rotated so as to view a built-in, fixed temperature black body reference source on the control unit to enable day to day calibration checks and adjustments to be made.

2.5 Steady State Technique

With the radiation pyrometer viewing the specimen in the direction normal to its surface, measurements of radiation pyrometer output signal, specimen temperature and ambient temperature were made at regular intervals of specimen temperature between 400°C and 675°C in the case of the first specimen (Test Series A), and between 150°C and 500°C in the case of the second one (Test Series B). At particular temperatures within these ranges the angle of view was varied from 0 to 70° keeping the specimen temperature constant. Values of emissivity were determined from these measurements by the method described below in Section 2.7.

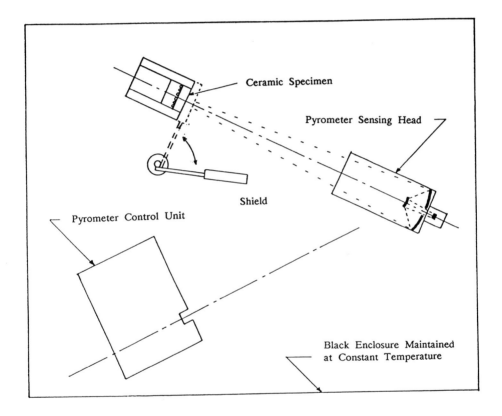

Figure 3.   Experimental Arrangement for Direct Measurements
of Emissivity (Steady State and Transient)

2.6 The Transient Technique

The special feature of the arrangement used in this case was a heat shield mounted on a rotating arm as shown in Figure 3. The shield consisted of a stainless steel plate of diameter 50 mm backed by insulating material. The shield was designed to fit over the surface of the specimen without touching it held in position by a spring. This insulated it from the surroundings and caused it to take up a uniform temperature. In order to ensure uniformity of temperature when operating at high values of specimen temperature, the shield could be separately heated by electrical means and its temperature could be monitored by chromel alumel thermocouples attached to it.

The arrangement was designed so that when the specimen reached thermal equilibrium at the required uniform temperature the shield could then be quickly swung away on the rotating arm, thus exposing the specimen to the surroundings and allowing radiant heat transfer to take place.   At this instant sampling of the radiation pyrometer output signal and the specimen and ambient temperatures commenced using a computer–based data acquisition system operating through an analogue to digital converter at high frequency.   Measurements were terminated after about 15 to 20 seconds.   Figure 4 illustrates the manner in which the pyrometer signal varied with time during such a transient.   By extrapolating the pyrometer output signal back to the start of the transient it was possible to determine the value corresponding to the initial temperature of the surface.   This was, of course, the temperature indicated by the thermocouples embedded below the surface because the initial condition was one of uniform temperature.

Such experiments were carried out at regular intervals of specimen temperature with the pyrometer viewing the specimen in the direction normal to its surface.   The temperatures ranges covered were 700°C to 950°C in the case of the first specimen (Test Series A) and 150°C to 1050°C in the case of the second one (Test Series B). Again, at particular temperatures within these ranges, experiments were carried out with the pyrometer viewing the specimen at various angles from 0 to 70° keeping the temperature constant.   Values of emissivity were determined from the above measurements using the method of analysis described in Section 2.7, below.

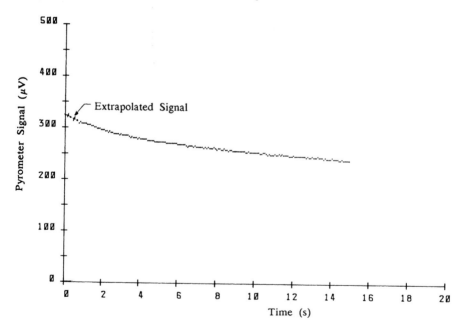

Figure 4.   Radiation Pyrometer Output Voltage versus Time for
                Typical Transient Test

2.7 Theoretical Basis of the Direct Method

As indicated above, the arrangement for the 'direct' method of emissivity measurement consists of a small heated specimen having a surface temperature T_s within a large black enclosure which is at a uniform temperature T_a close to the ambient value. The specimen is assumed to be opaque to thermal radiation.

A calibrated total radiation pyrometer views a small target area on the specimen surface in the normal direction or at some other chosen angle θ to the normal. The intensity of radiation received by the instrument is composed of two parts, one being that which has been emitted by the specimen and the other being that part of the black body radiation from the surroundings which is incident on the specimen surface and is reflected into the pyrometer by it.

Thus

$$i(\theta) = \frac{1}{\pi} \int_o^\infty \epsilon_\lambda(\theta, T_s) e_{b_\lambda}(T_a) d\lambda$$

$$+ \frac{1}{\pi} \int_o^\infty \rho_\lambda(\theta, T) e_{b_\lambda}(T_a) d\lambda \tag{1}$$

The black body radiation from the surroundings incident on the specimen surface is simply either absorbed or reflected in the case of an opaque material so it follows that

$$\alpha_\lambda(\theta, T_s) + \rho_\lambda(\theta, T_s) = 1 \tag{2}$$

Noting that according to Kirchoff's Law applied to spectral values of absorbtivity and emissivity

$$\alpha_\lambda(\theta, T_s) = \epsilon_\lambda(\theta, T_s) \tag{3}$$

so it follows that

$$\rho_\lambda(\theta, T) = 1 - \epsilon_\lambda(\theta, T_s) \tag{4}$$

Hence equation 1 can be re-written as

$$i(\theta) = \frac{1}{\pi} \int_o^\infty \epsilon_\lambda(\theta, T_s) e_{b_\lambda}(T_s) d\lambda$$

$$+ \frac{1}{\pi} \int_o^\infty [1 - \epsilon_\lambda(\theta, T_s)] e_{b_\lambda}(T_a) d\lambda \tag{5}$$

On the assumption that the spectral distribution of emissivity is independent of temperature, we replace $\epsilon_\lambda(\theta, T_s)$ in the second integral of equation 5 by $\epsilon_\lambda(\theta, T_a)$ and then noting that the total emissivities at temperatures T_s and T_a are respectively defined by

$$\bar{\epsilon}(\theta, T_s) = \frac{1}{\sigma T_s^4} \int_o^\infty \epsilon_\lambda(\theta, T_s) e_{b_\lambda}(T_s) d\lambda \tag{6}$$

and

$$\overline{\epsilon}(\theta, T_a) = \frac{1}{\sigma T_a^{\;4}} \int_o^\infty \epsilon_\lambda(\theta, T_a) \; e_{b_\lambda}(T_a) d\lambda \qquad (7)$$

equation 5 becomes

$$i(\theta) = \overline{\epsilon}(\theta, T_s) \frac{\sigma T_s^{\;4}}{\pi} + [1 - \overline{\epsilon}(\theta, T_a)] \frac{\sigma T_a^{\;4}}{\pi} \qquad (8)$$

If, in equation 8, we approximate $\overline{\epsilon}(\theta, T_a)$ by $\overline{\epsilon}(\theta, T_s)$ we obtain

$$\overline{\epsilon}(\theta, T_s) = \frac{i(\theta) - \dfrac{\sigma T_a^{\;4}}{\pi}}{\dfrac{\sigma T_s^{\;4}}{\pi} - \dfrac{\sigma T_a^{\;4}}{\pi}} \qquad (9)$$

Knowing the enclosure temperature T_a, the surface temperature T_s (obtained in the case of the steady state technique by carefully extrapolating the measured temperature distribution within the material back to the surface), and the intensity of radiation $i(\theta)$ (obtained by converting the output signal to intensity using the calibration equations) equation 9 can be used to evaluate the emissivity $\overline{\epsilon}(\theta, T_s)$.

The approximation involved in the above procedure can be shown to introduce an error of less than 1% for the conditions of the present experiments ($T_s \geqslant 150\,^\circ C$, $T_a \simeq 20\,^\circ C$).

In principle, once the emissivity has been found as a function of specimen temperature using the above procedure, $\overline{\epsilon}(\theta, T_a)$ can be found by extrapolation and then equation 8 can be used to determine improved values for $\overline{\epsilon}(\theta, T_s)$. However, in practice such refinement is difficult to justify in view of the very small error involved in the initial approximation, which is negligible compared with the uncertainties associated with experimental error.

3.   'INDIRECT' MEASUREMENTS OF EMISSIVITY

3.1   The Method

The 'indirect' method of determining the emissivity of a material is also very simple and has been widely used.   A black body radiation field of known intensity is produced inside a closed cavity by heating its walls so as to maintain them at a uniform temperature.   An unheated specimen, initially at ambient temperature, is rapidly introduced through a small hole into the cavity on a traversable arm.   A calibrated radiometer situated outside the cavity views the specimen and measures the emitted and reflected radiation emanating from it.

Knowing the intensity of the black body radiation field within the cavity and the temperature of the specimen, the emissivity can be deduced invoking Kirchoff's Law.

3.2 The Test Specimens

The specimens (Figure 5), of which there were several, were in the form of discs of diameter 25 mm and thickness 3 mm, 'fired' in thin–walled stainless steel cups.

Specimen temperature was measured by means of a 0.5 mm diameter stainless steel sheathed, chromel alumel thermocouple embedded in the refractory material with its junction close to the specimen surface.

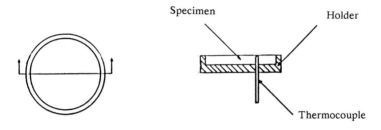

Figure 5.   Test Specimen Details for Indirect Measurements

3.3 Experimental Arrangement and Procedure

Figure 6 shows the experimental arrangement used for the 'indirect' measurements of emissivity.

The black body radiation cavity, which was produced by adapting a thick–walled mild steel pressure vessel, could be heated on the outside so as to maintain it at very uniform and steady known values of temperature up to 400 °C.

A calibrated total radiation pyrometer, mounted on a vertical traversing arrangement, had a rigid arm connected to it on which a test specimen could be attached inclined at some chosen angle to the optical axis of the pyrometer. The pyrometer was of the same type as that used in the direct measurements of emissivity which have been described earlier in this paper.

With the black body cavity temperature steady at the required value the pyrometer was focussed on a target area in the centre of the specimen and then the arrangement was traversed quickly so as to move the specimen into the black body cavity for a few seconds (to the position indicated by the dotted lines) and then out again. The pyrometer moved simultaneously with it but remained outside the cavity.

The pyrometer output signal and the test specimen thermocouple e.m.f. signal were sampled throughout at short intervals of time using a computer based data acquisition system and the results were processed 'on line' to determine the emissivity using the method described below.

Measurements were made with the black body cavity at 200°C, 300°C and 400°C.   At each of these temperatures, several nominally identical specimens were tested and in each case the angle of view was varied in steps of 10° from 25° to 65° to the normal to the surface of the specimen.

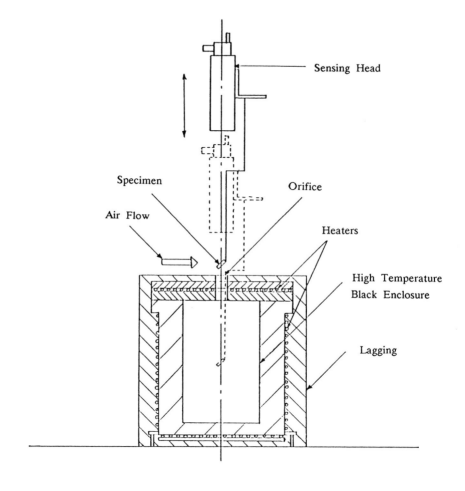

Figure 6.   Experimental Arrangement for Indirect Measurements

3.4   Theoretical Basis of the 'Indirect' Method

The arrangement under consideration here is a small unheated specimen (again assumed opaque to infra-red radiation) which is at a temperature T_s near to the ambient temperature and which is within a large black body cavity maintained at a steady uniform high temperature T_b.   A total radiation pyrometer situated outside the cavity views the specimen through a small hole at an angle θ to the normal to the specimen surface.

The intensity of radiation received by the instrument is again composed of two parts, one being that which is emitted by the specimen and one being that the fraction of the black body radiation field within the cavity which is reflected into the pyrometer by the specimen. Thus the theory presented earlier in Section 2.7 leading to equation 8 is also applicable here (except that T_a must be replaced by T_b). Thus we have

$$i(\theta) = \overline{\epsilon}(\theta, T_s) \frac{\sigma T_s^4}{\pi} + \left[1 - \overline{\epsilon}(\theta, T_b) \right] \frac{\sigma T_b^4}{\pi} \qquad (10)$$

If, we approximate $\overline{\epsilon}(\theta, T_s)$ by $\overline{\epsilon}(\theta, T_b)$ in this equation we obtain

$$\overline{\epsilon}(\theta, T_b) = \frac{\dfrac{\sigma T_b^4}{\pi} - i(\theta)}{\dfrac{\sigma T_b^4}{\pi} - \dfrac{\sigma T_s^4}{\pi}} \qquad (11)$$

which, knowing the specimen temperature T_s, the black body cavity temperature T_b and the intensity of radiation $i(\theta)$, can be used to evaluate $\overline{\epsilon}(\theta, T_b)$. If this is done for a series of experiments in the course of which T_b is varied, the resulting distribution of $\overline{\epsilon}(\theta, T_b)$ can be extrapolated to obtain an estimate of $\epsilon(\theta, T_s)$. This can then be used in conjunction with equation 11 to re-evaluate $\overline{\epsilon}(\theta, T_b)$.

4. RESULTS AND DISCUSSION

4.1 Presentation of Results

The results are presented are in the form of:

(a) A plot of emissivity as a function of temperature of the data obtained using the three different measurement techniques for both of the test series (Figure 7).

(b) Polar plots showing typical results obtained for directional emissivity as a function of angle of view using each of the three measurement techniques (Figure 8).

4.2 Discussion of Results

As mentioned earlier in the introduction to this paper, severe practical difficulties are encountered in the determination of the emissivity of ceramic materials, the greatest of which is undoubtedly the accurate measurement of specimen surface temperature. It was this aspect of the problem which provided the incentive for the present comparative study of three different techniques for the measurement of total emissivity.

The most obvious approach namely, that of 'direct' measurement of the radiant energy from a heated specimen emitting energy under steady state conditions, involves extrapolating the temperatures indicated by thermocouples embedded within the specimen at various depths below its surface, with all the uncertainties that this entails.

The alternative approach of making 'direct' measurements of emitted radiation under transient conditions, starting with a specimen which is initially completely shielded from the surroundings and is then suddenly exposed, again involves extrapolation, but this time of output signal from the radiation pyrometer. One problem here is that this signal inevitably has a significant amount of low frequency noise superimposed on it. Another problem is that in the event of the material being to some extent semi-transparent to thermal radiation (a known characteristic of ceramic materials operating at very high temperature) questions arise as to the extent to which an emissivity measured this way will be the same as that measured by other methods.

Similar questions arise with the 'indirect' approach, in which an unheated specimen reflects black body radiation from heated surroundings into a radiation pyrometer. Furthermore, the problem of surface temperature measurement is present again in this case (very steep gradients of temperature rapidly develop near the surface of the specimen as it absorbs heat from the surroundings).

In view of these various difficulties, and others which have not been mentioned, it is very pleasing that the values of emissivity yielded by the three methods employed here, in two separate series of tests on a variety of specimens, are in such excellent agreement with each other.   As can be seen by examination of Figure 7, a very well-defined distribution of emissivity as a function of temperature has been established from 150°C to 1000°C for this refractory material.   Measurements of the directional emissivity in all cases show little variation of emissivity with angle of view (Figure 8), confirming, as expected, that the material is effectively a diffuse emitter.

Assessments of uncertainty and error in measurements of the kind reported here are always difficult to make.   However, an attempt has been made to do this in the case of all three techniques and it is concluded that a curve fit to the experimental results presented in Figure 6 would describe the emissivity of this material to an accuracy of better than ± 8% over the entire range of temperature covered.

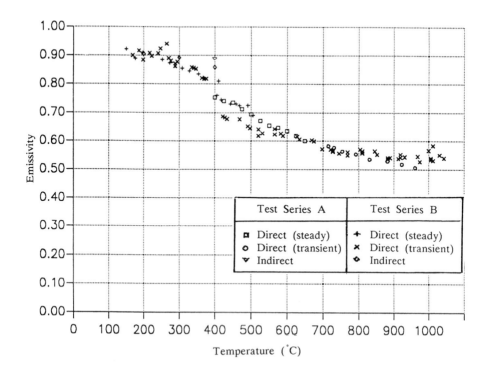

Figure 7.   Total Normal Emissivity as a Function of Temperature

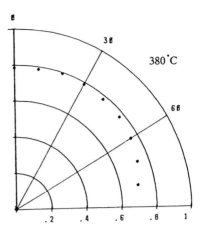

b)   Direct Method – Steady

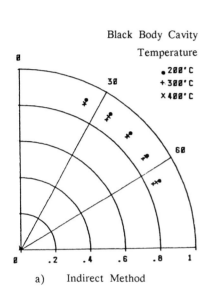

a)   Indirect Method

Figure 8.   Directional Emissivity

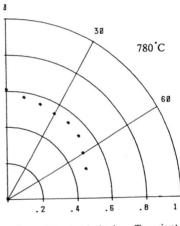

c)   Direct Method – Transient

4.3 Spectral Emissivity Measurements

The reduction of emissivity with increase of temperature observed in the present study is consistent with distributions of spectral emissivity of the kind which have been reported in the literature for similar materials.   However, it should be pointed out that such data are sparse and probably of questionable accuracy.   Thus, in parallel with the present study of total emissivity, work has been in progress in the authors' laboratory to develop equipment for making spectral measurements.   This equipment has now been successfully commissioned and a programme of experimental work is in progress to study the distribution of the emissivity with wavelength of the refractory material reported on here.   The method being used in these spectral measurements is a 'direct' one similar to that outlined in Section 2 of this paper.   Infra-red radiation from a target area on a heated specimen, or from a standard black body reference source, passes through a chopper and is focussed onto the entrance slit of a monochromator by a specially designed imaging optical system.   Inside the monochromator, mirrors collimate the infra-red beam and direct it to a diffraction grating which is mounted on a turntable.   For a given angular setting of the grating the first and higher orders of diffraction give rise to radiation with discrete wavelengths being directed via a focussing mirror and a suitable high pass filter to the exit slit of the monochromator, which allows a particular narrow waveband of the lower order radiation through. This radiation is projected by a small zinc selenide lens onto a pyroelectric detector which supplies an electrical signal through high pass and low pass filters to a lock-in amplifier. The equipment is producing excellent results which will be reported separately in the near future.

4.4 High Emissivity Coatings

As mentioned earlier in this paper, our studies of ceramic emissivity were initiated as a result of interest in evaluating the effectiveness of so-called 'high emissivity coatings' used to enhance the emissivity of furnace refractory linings. The results obtained to date in the course of our studies of such coatings have exhibited some interesting features. Application of one such coating actually brought about a significant decrease in the total emissivity of the refractory material for temperatures below 600°C, followed by a progressive increase beyond that temperature. At 1000°C the enhancement of emissivity was about 20% and the indication was that emissivity would continue to increase with further rise of temperature. Our present studies of spectral emissivity are aimed at providing a better understanding of these effects and ultimately enabling reliable predictions to be made of the likely effectiveness of such coatings.

5. CONCLUSIONS

The three methods of measurement which have been compared in the present studies have been yielded very consistent results when used to establish the variation of total emissivity of a refractory material (composition 62.3% Al_2O_3, 32.1% SiO_2) over the temperature range 150°C to 1000°C.

The emissivity is about 0.9 at 150°C and it decreases with increase of temperature, slowly at first but more rapidly beyond 300°C, reaching a value of about 0.6 at 650°C. Beyond this temperature the emissivity levels out, reaching about 0.55 at 800°C. Directional measurements have exhibited little variation of emissivity with angle of view in the range 0 to 70° to the normal, indicating that the material is a diffuse emitter.

ACKNOWLEDGEMENT

The authors gratefully acknowledge the advice, practical help and encouragement provided in the course of the investigation by Mr. R.J. Tucker of the Midlands Research Station of British Gas Plc.

REFERENCES

Cabannes F 1970 *High Temperatures-High Pressure* **8** pp 155-166
Dunkle R V and Gier J T 1953 *Progress Report* University of California Berkeley
Fiveland W A 1985 *Institute of Energy Workshop on the Emissivity of Furnace Materials* British Gas Corporation Midlands Research Station, England
Fletcher J D and Williams A 1984 *Journal of the Institute of Energy* pp 377-380
Gubareff G G, Janssen J E and Torborg R H 1960 *Thermal Radiation Properties Survey* (Mineapolis: Honey Regulator Company)
Hass G 1964 *Symposium on Thermal Radiation of Solids* NASA Sp-55
Hobbs H A and Folweiler R C 1966 *Technical Report ASD-TDR-62-719* US Air Force
Pears C D 1962 *Progress in International Research in Thermal and Transport Properties* **8 54** pp 589-597
Singham J R 1962 *International Journal of Heat and Mass Transfer* **5** pp 67-76
Touloukian Y S 1970 *Thermophysical Properties of Matter TRGP series 8* (Plenum)
Wade W D and Slemp W S 1982 NASA TN D-998
Whitson M E 1976 *Handbook of Infrared Properties* NASA report N76-14275

Spectral emittance measurements of furnace wall materials and coatings

E. Hampartsoumian

Department of Fuel and Energy
The University of Leeds, Leeds LS2 9JT

ABSTRACT: An   accurate   knowledge   of   the   effective   emissivity,
absorptivity and reflectivity of a surface is an essential requirement
for   the   evaluation   of   high   temperature   furnace   heat   transfer
processes.   This paper describes new experimental data on the spectral
and total normal emittance of modern furnace insulating refractories
and coatings at high temperatures for which little data has previously
been available.

1 INTRODUCTION

In recent years, there has been a considerable growth in the use of
new   lightweight   insulating   ceramic   linings   and   surface   coatings   in
furnaces.   The cost saving benefits have usually been stated to include:

i)   the   prevention   of   surface   degradation   by   chemical   attack,   thermal
stress and abrasion.

ii)   the   reduction   of   heat   losses   and   enhancement   of   the   thermal   radiation
properties of the furnace enclosure.

The   former   serves   to   prolong   the   useful   life   of   a   furnace   whilst   the
latter   are   intended   to   lead   to   a   reduction   in   fuel   usage   for   a   particular
throughput.   In   this   context,   the   use   of   high   emissivity   coatings   have
often   been   advocated   to   lead   to   beneficially   higher   rates   of   heat   transfer
to   the   stock   (Lasby   1982,   Fisher   1989).   In   view   of   the   claims   and
uncertainties   regarding   the   performance   of   these   and   other   novel   furnace
materials,   there   is   a   need   for   information   on   the   radiative   properties   of
such   materials   in   order   to   be   able   to   conduct   reliable   predictions   of
furnace performance.

The   emissivity   of   a   surface   is   an   important   radiation   property   for
heat   transfer   analysis   and   for   temperature   measurement   by   radiation
thermometry.   The   values   most   commonly   used   or   assumed   in   furnace
applications   are   usually   the   simplest   which   relate   to   radiation   at   all
wavelengths   and   in   all   directions   (total,   hemispherical   emissivity).   Most
furnace   materials   of   practical   interest   exhibit   non-grey   behaviour   so   it
is   also   necessary   sometimes   to   know   the   spectral   emissivity
characteristics   of   surfaces.   The   spectral   region   of   interest
corresponding   to   typical   furnace   operating   temperatures   is   the   infra-red

region beyond 1.5μm and up to 30μm.   The proportion of the blackbody spectrum above 35μm is negligible; that below 1μm is less than 0.8% at 1400K.   The term emissivity is normally reserved for ideal homogeneous materials which are characterized by perfectly smooth surfaces.   Emittance is used for the more typical non-ideal materials, as used for example in furnace construction, where the radiative properties are not unique properties of the bulk material but may also depend on other characteristics such as surface finish.   Most industrial and research workers however usually refer to the emissivity of a material and this terminology will also be used throughout this paper.

The present paper describes work on the measurement of the spectral emissivities of a range of modern furnace materials.   This work is part of an ongoing research programme looking at the importance of wall and stock emissivity on furnace heat transfer processes (Alexander *et al* 1988, Gray *et al* 1989).

2. DEFINITIONS OF RADIATIVE PROPERTIES OF SURFACES

The emissivity of a surface may be defined as the ratio of the radiance from the surface to that from a blackbody viewed under identical optical and geometrical conditions and at the same temperature.   The total blackbody radiation flux density, or emissive power, $E_b(T)$ at a fixed temperature T is obtained from integration of the blackbody spectral flux $e_b(\lambda, T)$ over all wavelengths

$$E_b(T) = \int_{\lambda=0}^{\infty} e_b(\lambda, T) d\lambda \tag{1}$$

This integral has been evaluated to yield the familiar expression

$$E_b(T) = \sigma T^4 \tag{2}$$

Based on the definition of emissivity, the total temperature dependent emissivity of a surface, $\epsilon_t(T)$, may be written in terms of the blackbody emissive power and spectral flux density of the surface, $e(\lambda, T)$, as

$$\epsilon_t(T) = \frac{1}{E_b(T)} \int_{\lambda=0}^{\infty} e(\lambda, T) d\lambda \tag{3}$$

Integration of $e(\lambda, T)$ over the spectral band limits (λ_1, λ_2) will yield instead the spectral emissivity $\epsilon_{\lambda j}(T)$

$$\epsilon_{\lambda j}(T) = \frac{1}{\Delta E_{b\lambda j}(T)} \int_{\lambda_1}^{\lambda_2} e(\lambda, T) d\lambda \tag{4}$$

where λ_j is the nominal wavelength position defined by

$$\lambda_j = (\lambda_1 + \lambda_2)/2 \tag{5}$$

and $\Delta E_{b\lambda j}(T)$ is the blackbody band emissive power over the same wavelength range.

3. EMISSIVITY MEASUREMENTS

In general, the emissivities of real materials based on equation 4 can only be obtained by experiment. The most common methods are based on ratio-radiometric techniques where the radiance from a sample, sensed as a millivolt output using a suitable detector, is divided by the output produced from the same detector when viewing a suitable reference blackbody source at an identical temperature as the sample. The main experimental constraints are ensuring isothermal sample/blackbody condition and avoiding reflected radiation from the target surroundings into the detector field of view.

Since emissivity is a function of temperature as well as wavelength, to obtain emissivities of refractory materials at typical furnace temperatures requires that the samples are heated. A suitable system must be capable of attaining surface temperatures in excess of 1000K to correspond with realistic furnace operating conditions whilst at the same time not interfering with the measurement of the radiation emitted by the surface. Other complications when making measurements on refractory materials arise from their low thermal conductivity, irregular porous surfaces and translucent (at high temperature) behaviour. This leads to errors in true surface temperature measurement since the material will emit and absorb energy within a surface layer of finite thickness over which a steep temperature gradient may be present making it difficult to assign a unique temperature value to the reference blackbody. Several experimental systems were developed in order to carry out the measurements. The principal systems used are briefly described below together with their applicability.

3.1 Blackbody Comparison Technique

The experimental system used to measure emissivity is illustrated in Fig. 1. Two methods of sample heating were employed. In the first, the

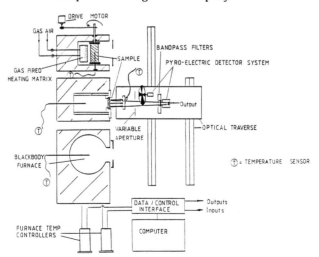

Fig. 1. Emissivity measurement system and ancillaries.

surface of a rotating cylindrical test sample was alternatively heated and
viewed by an optical system which could be traversed to receive radiation
either from the sample or from the reference blackbody. In the second
system the sample was back heated in order to represent a hot refractory
in an indirectly fired mode. A pyroelectric detector sensitive to
radiation between 0.2 and 24μm was used to make spectral measurements with
a series of broad bandpass interference filters. The spectral
transmission characteristics of the filters used are shown in Fig. 2. A
longpass filter was used to cover the spectrum above 7.3μm. Inevitably
the wings of the filters overlap and some spectral regions are not
adequately covered although for the purpose of the presentation of the
results top hat transmission characteristics with no overlaps or gaps
using the nominal bands shown in Fig. 2 have been assumed. All readings
were taken normal to the sample surface using a narrow optical aperture
giving a target diameter less than 5mm on the surface. Any effects of
surface irregularities or property variations are averaged by the system
as the sample rotates in the optical field of view. It can be shown that
for nearly all non-metallics, the ratio of hemispherical to normal
emissivity is within 5% of unity (De Witt and Hernicz 1972).

The ratio of the detector signal outputs, R, when identically viewing the
target, S_t, and the reference blackbody, S_b, is of the form,

$$R = \frac{S_t}{S_b} = \frac{\epsilon_{\lambda j} E_{b\lambda j}(T) + (1 - \epsilon_{\lambda j}) E_{b\lambda j}(T_s) + Z}{E_{b\lambda j}(T_b) + Z} \qquad (6)$$

In this relationship the second term of the numerator represents the
reflected portion of the irradiation from the surroundings and the third
term the background radiation in the system at a temperature T_s. If it is
assumed that the reference blackbody temperature, T_b and target
temperature, T, are equal and much greater than T_s, the measurement
equation for sample emissivity based on equation 6 reduces to the simpler
form

$$\epsilon_{\lambda j} = \frac{S_t - Z}{S_b - Z} \qquad (7)$$

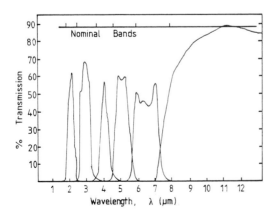

Fig. 2. Spectral trans-
mission characteristics
(percent) of bandpass
filters used in narrow
angle radiometer.

Table I. Composition and total emissivities of materials investigated.

Sample Type	Commercial Name	Typical Composition (%)							T Range (K)	ε TOTAL at T Extremes
		Al_2O_3	SiO_2	Fe_2O_3	MgO	TiO_2	ZrO_2	CaO		
Lightweight Refractories	MPK 110	20	53	4.0	16.4	1.2	–	1.4	1073–1300	0.58–0.62*
	MPK 125	36	46	1.1	0.4	0.2	–	14.6	1063–1350	0.54–0.64
	MPK 140	41	54	1.3	0.3	0.6	–	0.3	1073–1350	0.48–0.55
	MPK 155HA	61	36	0.4	0.1	0.1	–	0.2	1063–1350	0.44–0.36*
	MPK 130HSR	36	54	1.7	5.8	0.5	–	0.3	1073–1250	0.6 –0.62
	MPK SUPRA	11	75	6.0	2.0	1.0	–	0.5	1073–1300	0.64–0.81
Ceramic Fibres	KAOWOOL (Standard)	43–47	53–57	–	–	–	–	–	923–1173	0.87–0.40*
	ZIRCAR	–	–	–	–	–	99	–	1080	0.51
	SAFFIL	95	5	–	–	–	–	–	1010	0.49
	MICROTHERM (Standard)	2.4	64.5	0.3	–	31.9	0.2	0.04	950	0.48
"High Emissivity" Coatings	Carborundum Powder based							–	1023–1273	0.95–0.8*
	SiC based							–	950–1373	0.84–0.81
	ZrO_2 based							–	1073–1350	0.55–0.64
Protective Coatings	NONVIT	1.6	32.1	–	0.3	0.2	62.3	1.4	1073–1373	0.56–0.44

* See Figs. 3, 4, 5 for spectral properties

The spectral band emissivities at around 1073K obtained by the blackbody comparison technique of three representative materials from Table I are shown in Fig. 3. A significant general feature of all the dielectric materials investigated was a region of lower spectral emissivities at the shorter wavelengths (<4μm) followed by a region of high emissivities at longer wavelengths. This causes the total emissivity to be weighted towards the long wavelength spectral emissivities at low temperatures and the shorter wavelength emissivities at higher temperatures. Since the distribution of thermal radiation is biased more towards the shorter

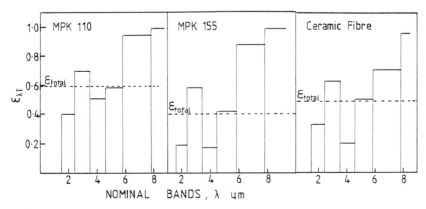

Fig. 3. Comparison of typical spectral band emissivities (at ~ 1073 K) for three materials as shown.

The magnitude of the background zero, Z, is obtained by identically viewing a cold source. Errors in measurement will mainly arise, as previously stated, from non-isothermal conditions between the target and reference blackbody. Such errors becoming more significant at shorter wavelengths and lower temperatures. A small correction can easily be made if the emissivity of the reference blackbody is less than unity.

The majority of the experimental results presented were obtained from the rotating sample system since true surface temperatures for the back-heated system proved difficult to establish due to large temperature gradients near the surface.

3.2 Reflectivity technique

An alternative approach is to measure the reflectivity (total or spectral) of a sample and to then infer the apparent emissivity by invoking Kirchoff's Law. This can be accomplished without a knowledge of the true temperature of the sample surface which is only momentarily heated. If it is assumed that the spectral emissivity properties of a surface does not change significantly when the surface temperature is varied, spectral properties measured at room temperature can be used to calculate the total emissivity at higher or lower temperatures. The detector signal output obtained by viewing a reference blackbody (S_b) is followed by a measurement from the target sample placed momentarily in the blackbody furnace (S_{t1}) and finally a reading from the sample outside the furnace (S_{t2}). The first measurement will be proportional to the blackbody emissive power; the second will comprise of the same components in the denominator of equation 6. Assuming that background zero is negligible, the apparent emissivity will thus be given by the ratio of the outputs as

$$\epsilon_\lambda = 1 - \left(\frac{S_{t1} - S_{t2}}{S_b} \right) \qquad (8)$$

This useful technique is quite commonly used for the routine analysis of industrial samples in order to obtain values of total emissivity for optical thermometry purposes. Errors may however arise from this technique due to the non-greyness of the test material and differences between the source and sample temperatures.

4. EXPERIMENTAL RESULTS AND DISCUSSION

4.1 Total and Spectral Emissivities

The total and spectral band emissivities of fourteen typical commercially available refractories, insulating materials and coatings were measured using the blackbody comparison technique for different surface temperatures in the range 900 - 1400K depending on the material. The materials investigated covered a wide range of compositions and physical properties as shown in Table I. The difficulties of measuring the true surface temperatures of the irregular refractory surfaces and in particular the ceramic fibres limits the accuracy of the data presented in Table I to around 15%.

wavelengths at furnace operating temperatures, the spectral emissivities in this region are of greater importance as regards heat transfer.

Some typical spectral emissivity trends obtained from the reflectivity technique are shown in Fig. 4. Similar trends to the results obtained by blackbody comparison may be seen although in the former case the sample is not exposed to the high temperature source for an adequate period of time for isothermal conditions to be established throughout it's volume although it is likely that a finite surface layer of molecules will be sufficiently excited to higher energy levels leading to the emission of thermal radiation.

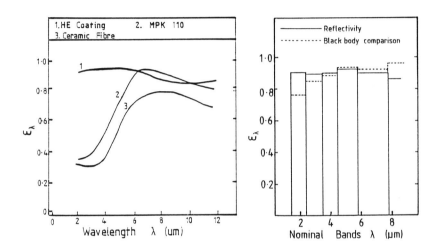

Fig. 4. Spectral emissivities obtained by reflectivity technique.

Fig. 5. Spectral band emissivities for a high emissivity coating.

Figure 5 shows a comparison between the experimental spectral band emissivity data obtained by the two techniques for a typical high emissivity coating. The coatings based on silicon carbide and carborundum (see Table I) showed spectral emissivities above 0.7 for all the bands measured. Not all the coatings investigated performed as well in this respect.

4.2 Effect of temperature on total emissivity

The effect of temperature is mainly determined by the behaviour of the emission (or absorption) bands of the material at the shorter wavelengths since the spectral energy distribution of the bands at the longer wavelengths was found to remain relatively fixed with respect to each other as the temperature was increased. This follows because the region of maximum energy of the Planck's function moves towards the shorter wavelengths as the temperature is increased and hence ϵ_t will decrease with increasing temperature if the shorter wavelength low emissivity bands remain relatively fixed. With the exception of one of the ceramic fibres, most of the materials showed only a small variation in ϵ_t over the relatively small range of surface temperatures investigated as shown in Fig. 6. Firm conclusions cannot be drawn for the former since potential errors in temperature measurement were compounded by the fibrous and translucent nature of this material although there is some agreement with the earlier measurements of Fletcher and Williams (1984) using a different technique.   There is thus a need for reliable correlations of emissivity data with temperature in order to conduct performance calculations for transient furnace operation.

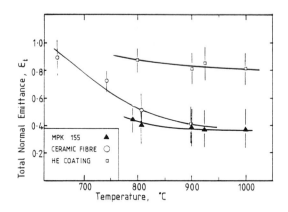

Fig. 6.   Variation of total normal emissivity with material surface temperature.

4.3 Effect of surface finish

The influence of surface roughness was investigated by either lapping the surface of commercial samples to vary the surface finish of by preparing samples from castable refractory with a range of surface finishes including, grooved, dimpled, and flat (rough and polished).   Only small differences were observed in emissivity which were within the limits of experimental error.   It is known that for metals wide variations in normal spectral emittance can stem from different surface conditions (De Witt and Hernicz 1972).   However, in this case the effect of oxidation or surface roughness is to change the characteristics of the metal from a metallic specular reflector (or emitter) to a more diffuse reflector.   The refractory materials investigated behave as diffuse emitters irrespective of the surface finish so that the geometric distribution of the radiant energy emitted or reflected from the surface is not likely to vary significantly with different angles of viewing up to around 70° form the normal.   Surface roughness is also likely to be less important if the material becomes semi-translucent at high temperatures since the radiative behaviour will then be dependent upon conditions within the bulk material below the surface.

5. CONCLUDING REMARKS

1.   The results presented have shown that the total normal emissivities of most common modern furnace refractories are typically in the range 0.4 to 0.6 at furnace temperatures.   Whilst there are no firm trends, of the samples investigated the lowest emissivities tended to result from the higher alumina materials designed to perform at higher temperatures.   In the case of furnaces where the emissivity of linings may be important (see Alexander *et al* 1988) the use of such lower emissivity materials could reduce performance.

2. Of the main experimental techniques employed, the blackbody comparison method can be considered to be more realistic. The reflectivity technique however provides a simpler means of comparing the spectral behaviour of different materials and does provide an acceptable level of accuracy.

3. In general the 'high emissivity' coatings tested lived up to their name although it should be remembered that the effects of long term exposure under realistic operating environments on the emissivity or useful life of the surface have not been established in these tests. The best high emissivity coatings may not therefore be the most durable.

6. REFERENCES

Alexander I, Gray W A, Hampartsoumian E and Taylor J M 1988 *Proc. 1st European Conf. on Industrial Furnaces and Boilers* (Lisbon: Infub)

De Witt D P and Hernicz R.S 1972 *Temperature, It's Measurement and Control in Science and Industry* 4 1 (Reinhold) pp459-482

Fisher G 1986 *Ceramic Bulletin* 65 pp283-287

Fletcher J D and Williams A 1984 *J.Inst. Energy* 58 pp 377 - 380

Gray W A, Hampartsoumian E, Taylor J M and Williams A 1989 *Proc. 1989 Int. Gas Research Conf.*

Lasby S B 1982 *Industrial Heating* 49 pp 49-52

Acknowledgements

The author would like to thank the SERC for a grant in aid of this work and to the Wolfson Combustion Instrumentation and Control Unit, University of Leeds, for assistance in carrying out the emissivity measurements.

The spectral emissivity of glass furnace roofs and its effect on heat transfer

J A Wieringa, J J Ph Elich, C J Hoogendoorn[1]

Delft University of Technology, Faculty of Applied Physics, P.O.Box 5046, 2600 GA Delft, The Netherlands

ABSTRACT The influence of the roof emissivity on the radiative heat flux or fuel consumption in (regenerative) glass furnaces was studied using a well-stirred furnace model and spectral optical properties of gas and refractory. The importance of the gas spectrum was considered. In a regenerative furnace, increasing the roof emissivity to 0.95 was found to decrease the fuel consumption by 2 to 3 %. Without a regenerator, a flux enhancement of 3 to 4 % was found. In smaller furnaces these values will be higher. The emissivity of flat and grooved refractory material were determined.

1. INTRODUCTION

In the last several years environmental problems like the greenhouse effect and acid rains give a strong impulse to decrease the consumption of fossil fuel and the emission of NO_x. Since glass melting furnaces are known for their high gas temperatures (and consequently NO_x rates) and their high use of energy, it is worth to improve the production process. In this study the accent is on radiative heat transfer in the combustion chamber in connection with the radiative properties of the roof. Because of the high flame temperatures in glass melting furnaces, radiative heat transfer is much more important than convective heat transfer.

Several authors (Gosman *et al* 1980, Carvalho *et al* 1984, Post 1987, 1988, Post and Hoogendoorn 1988) published about simulations of glass furnace combustion chambers, but their work mainly concerned the flow and combustion in the combustion chamber, and did not consider spectral radiation effects. Since 1985 the effects of the non-greyness of the combustion gases in natural gas fired steel furnaces were discussed by Tucker (1985), Docherty and Tucker (1986a, 1986b), Elliston *et al* (1987) and Alexander *et al* (1988). They found that in continuously operating furnaces with not too high wall heat losses, a high refractory emissivity promotes the heat transfer to the furnace load.

The refractory emissivity has influence on heat transfer due to the non-greyness of the combustion gases in the flame. The radiation from the flame is to a great extent limited to several wavelength bands. In fact these bands consist of thousands of lines, emanating from the energy levels of the molecules. Hottel and Sarofim (1967) showed that in non-luminous flames, radiation can be characterized accurately by acounting only for the gases H_2O and C_2O. If the roof emissivity is low, most of the flame radiation is reflected and the wavelength distribution

[1]This study was performed under contract with the Netherlands Agency for Energy and Environment (NOVEM)

remains unchanged. Passing through the emission/absorption bands of the flame again, most of the radiation is absorbed in the flame and only little of the reflected heat flux will reach the glass through the flame. However, when the refractory emissivity is high, most of the radiation is absorbed in the roof and will be reradiated according to the Planck black body spectrum appropriate to the roof temperature. Then a large part of the re-emitted radiation goes through the clear "windows" in the flame spectrum.

Docherty and Tucker first reported predictions of fuel savings due to this effect up to about 30 %. Later the predictions became less optimistic: Elliston found a maximum flux enhancement of 5 % and also Alexander *et al* mention at most 10 %. All these results refer to non-regenerative furnaces, operating at considerably lower temperatures than glass furnaces.

In this work the influence of the roof emissivity is studied for (regenerative) glass furnaces. First the influence of several parameters is shown using a 15 bands gas spectrum from the work by Alexander *et al* and Elliston *et al*. Next, a more detailed spectrum is used, that has been computed for the correct geometry and gas composition. The spectral emissivity of the refractory has been measured and this spectrum is used in the computations. Furthermore, attention will be given to possible ways to increase the refractory emissivity.

2. THE FURNACE

The modelled furnace is a regenerative, natural gas fired, continuously operating glass melting furnace with three burner ports on each side of the glass melt that are cyclically fired. Each burner port contains two burners, that consist of a nozzle injecting the fuel and primary air into a stream of preheated secondary air. Because of symmetry, it is only necessary to simulate half a burner compartment, as shown in Figure 1, that has the following dimensions: length 7.3 m, average height 2.1 m and width 1.7 m.

The mass flow of 0.059 kg/s natural gas (lower calorific value 44.6 MJ/kg) is burned with 10 % excess air. The burnout has been taken 96 % (Post 1988). The secondary air is preheated in the regenerator giving a temperature of 1150 °C. This leads to a high adiabatic flame temperature of 2300 °C, which makes it necessary to take dissociation into account.

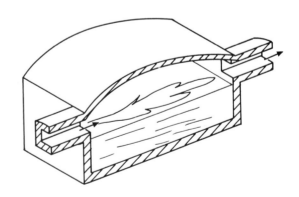

Fig.1. The combustion chamber

The glass temperature is taken 1500 °C, the glass emissivity is 0.8. Heat losses through the roof amount 94 kW with a heat transfer coefficient of 3.3 W/m^{2}K. The roof emissivity is 0.4 at the temperature of the roof (about 1850 K). This will be discussed in a later section.

3. THE RADIATION MODEL

The computer model that was used to predict radiative heat transfer in the glass furnace in this study is a well-stirred furnace model. If we assume a uniform temperature distribution in the flame and at the at the roof and the glass surface, and if we characterize the gas volume by a single mean beam length, we can use the network method (Oppenheim, 1956). Although the well-stirred furnace model is a rather simple computational method, the results were found to be accurate, especially for long flames, compared to the solution of a complete flow and combustion model (Post 1988).

Figure 2 shows the electrical analogue of the equations that hold for a grey gas, i.e. if the gas emissivity is taken independent of wavelength. The resistors in the network are found from gas and surface emissivities and the areas of glass and roof. The voltage sources represent the black body emissive powers of the relevant parts of the furnace chamber. For such a network the electrical currents symbolizing the radiative heat fluxes can be solved analytically.

However, since the temperature of the gas is not known, we have to apply a heat balance to the gases in the combustion chamber. That is:

$$Q_{air} + Q_{fuel} - Q_{glass} - Q_{roof} - Q_{flue\ gas} = 0 \ , \tag{1}$$

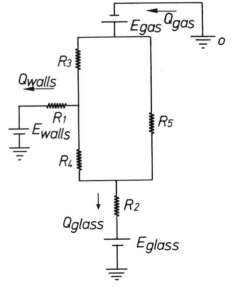

where Q represents a heat flow. Using this relation, we can solve the network and the heat balance iteratively.

The network in Figure 2 applies to a grey gas, but it can easily be extended to describe a spectral gas and/or refractory. Then, in a case of N spectral bands, a set of N of these networks has to be used. In each network the total black body emissive powers of the roof, glass and gas are substituted by the emissive powers E_n between the two wavelength limits of the spectral interval n:

$$E_n = \int_{\lambda_n}^{\lambda_{n+1}} \frac{c_1 \lambda^{-5}}{\exp\left[\frac{c_2}{\lambda T}\right] - 1} \ d\lambda \ . \tag{2}$$

Fig.2. Electrical analogue of the grey well-stirred furnace model

Here λ is wavelength, c_1 and c_2 are the first and second radiation constants, and T is the temperature of the roof, gas or glass. Within each spectral band the emissivity is assumed to be constant over wavelength. The spectral networks are coupled by the fact that the corresponding temperatures have to be the same in all these networks. A total flux can be found by summing the spectral fluxes over all wavelength bands.

The spectral model has been used in two ways. First the effect of increasing the roof emissivity has been studied, keeping the mass flows of fuel and air constant. This way the change in the heat flux to the glass melt is found. Here we also keep the secondary air temperature constant.

Secondly the heat flux to the glass is kept constant and the reduction of the fuel consumption is found. To keep the air factor constant, also the amounts of primary and secondary air are varied with the fuel flow. Here we also simulate the effect of the regenerator on the secondary air enthalpy, that is described in the next section. Especially this latter method will be of importance to glass industry, because there reduction of fuel consumption and flame temperature are aimed at instead of flux enhancement.

4. THE REGENERATOR

As stated before, increasing the refractory emissivity will improve the heat transfer from the flame to the load. From the heat balance, eqn. (1), a lower flue gas temperature will then be found. Thus the regenerator will transfer less heat to the secondary air, so that partly the heat transfer enhancement is compensated. When the heat flux to the glass is kept constant and the fuel mass flow is reduced, the gas temperature will drop even further.

This can be simulated in the computer model by putting:

$$Q_{sec.air} = \eta_{reg} \, Q_{flue \; gas} \qquad (3)$$

where $Q_{sec.air}$ and $Q_{flue \; gas}$ are the enthalpies of the secondary air and combustion gases, and η_{reg} is the regenerator efficiency. η_{reg} has been taken 0.44 and is assumed to be independent of the flue gas temperature. Eqn. (3) has been applied in the computations where the heat flux to the glass has been kept constant (except in one case that will be mentioned later).

5. THE GAS SPECTRUM

Two basically different gas spectra have been used in the heat transfer computations. Both are shown in Figure 3. The first has been taken from

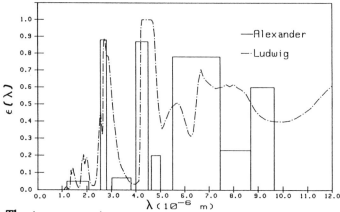

Fig.3. The two gasspectra

Alexander *et al* (1988) and consists of 15 spectral bands of which 7 are transparent. It was calculated according to data by Edwards (1960).

Although some results have been obtained with this spectrum, it has some shortcomings for our purposes: the bands in the spectrum do not reproduce the exact form of the absorption bands and furthermore it was calculated for different gas temperatures and for a smaller furnace. The (estimated) mean beam length of the geometry of Alexander *et al* is 1.4 m. Scaling to larger gas volumes has been done by finding the spectral absorption coefficients k_λ and then varying the mean beam length L in:

$$\varepsilon_\lambda = 1 - \exp(-k_\lambda L) \tag{4}$$

where ε_λ is the spectral emissivity. In our glass furnace geometry the mean beam length is 2.8 m.

The other two mentioned problems were reason to calculate a second spectrum according to data by Ludwig *et al* (1973), based on the statistical narrow band model (Goody, 1964). They give tables of experimental data on the spectral absorption coefficients and line densities of H_2O and CO_2. The line shape that has been used is the combined Lorentz-Doppler broadened line. The probability density function P of the line strength S of the spectral lines was taken $P(S) \propto S^{-1}$ for S $\leq S_m$ or, leading to the same results, $P(S) \propto S^{-1}\exp(S/S_m)$. Soufiani *et al* (1985) have shown that for Lorentz lines this distribution gives very accurate results. Pressures of H_2O and CO_2 have been found from flow computations by Post (1987, 1988) and these are 15 kPa for H_2O and 8 kPa for CO_2.

The resulting spectrum is also depicted in Figure 3. It consists of 372 spectral wavenumber intervals of 25 cm^{-1}, here converted to wavelength units. The two spectra show several differences, that may be explained by the different gas temperatures at which the spectra have been calculated, mean beam lengths and even gas compositions, and also by the different emissivity models. The results obtained according to Edwards are based on wide band properties, instead of the narrow band data from Ludwig *et al*.

Using the Ludwig spectrum, two computational experiments have been carried out. The first was to split up each 25 cm^{-1} interval in three smaller, equal intervals. In the middle one the emissivity was increased by an amplitude Δ; in the other two the emissivity was decreased by $\Delta/2$ as in Figure 4. Of course, care was taken that the gas emissivity always remained between 0 and 1. This way separate spectral lines have been simulated roughly, keeping the average gas emissivity very well constant. The other experiment was to increase the number of "lines" in each 25 cm^{-1} band at some value of Δ. Here the number of lines per band was always even, and alternatingly the emissivity change was equal to $+\Delta$ or $-\Delta$.

6. THE SPECTRAL PROPERTIES OF THE REFRACTORY

In order to estimate the enhancement of the furnace efficiency, it is necessary to know the (spectral) emissivity of the refractory material that is currently used in the glass melting furnaces. The most common material is silica, that mainly consists of SiO_2. If the emissivity is already high, only a small margin will be left to improve the efficiency.

The normal-hemispherical spectral reflectivity $\rho_n(\lambda)$ has been measured at room temperature using an integrating sphere spectrophotometer with a monochromator. The measurements were relative to the known reflectivity of gold, so that absorptive effects of the air along the radiative paths were eliminated. The normal spectral emissivity $\varepsilon_n(\lambda)$ is found from:

$$\varepsilon_n(\lambda) = 1 - \rho_n(\lambda) \ . \tag{5}$$

The measured spectrum is shown in   Figure 5. Experimental uncertainties are within 5 % of the measured values. Assuming that the measured spectral emissivities are independent of temperature (which has been confirmed by other measurements (private communication, 1988)), we can calculate the temperature dependence of the total emissivity. This dependence is an effect of the shift of the Planck distibution to smaller wavelengths with higher temperatures. It was found that at the temperature of the roof the emissivity drops to about 0.4. Then the maximum of the Planck distibution is at about 1.6 μm, where $\varepsilon_n(\lambda)$ is low.

The hemispherical emissivity $\varepsilon(\lambda)$ can be found from integration of the angular emissivity $\varepsilon_\vartheta(\lambda)$. Our measurements of this quantity show a strong

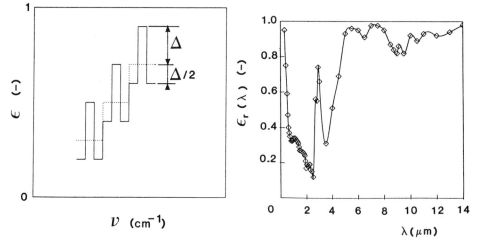

Fig.4. Superposition of the spectral "line" structure

Fig.5. The measured refractory spectrum

decrease for angles approaching 90°, which is a typical behaviour for insulators. Integration over the hemisphere shows that $\varepsilon(\lambda)$ is about 3 % lower than $\varepsilon_n(\lambda)$. This small effect has been neglected in further use of the measured spectrum, however.

7. RESULTS

The effect of increasing the roof emissivity on the heat flux to the load can be seen in Figure 6. In this simple case the 15 bands spectrum from Alexander *et al* (1988) was used and the regenerator and the spectral properties of the roof were not taken into account. In the figure distinction is made between the direct flux from the flame to the load and the flux that is either reflected or absorbed and re-emitted by the refractory. It can be seen that increasing the refractory emissivity stimulates the flux from the refractory. This causes the flame temperature to drop, and with it the direct flux. However, the net result is a heat transfer enhancement. In this case, increasing the refractory emissivity from 0.4 to 0.95 leads to an increased heat flux to the glass melt by 7.9 % (from 83.5 to 90.1 kW/m²). This result is comparable with the result of Alexander *et al.* The flame temperature is reduced by 51 K (to 2068 K). The roof temperature is raised somewhat, but not more than about 6 K.

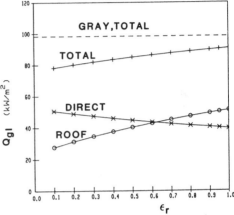

Fig.6. Effect of the roof emissity on heat transfer

If the roof emissivity is increased from the measured spectrum to 0.95 (grey), a heat flux enhancement of 5.4 % is found. Although this spectrum has an average emissivity of 0.4, it appears to give a larger flux than the grey roof with emissivity 0.4.

The regenerator, used while keeping the heat flux to the glass constant, leads to a fuel saving of 5.3 % when the emissivity is increased from 0.4 to 0.95 (both grey). Because of the lowering of the secondary air temperature, the flame temperature is now decreased by 60 K, and the roof temperature rises only by 1.2 K. For comparison: if the heat flux is kept constant without considering the regeneration effect, the reduction of the fuel consumption is found to be 9.4 %.

The effect of the combustion chamber volume was studied by increasing the mean beam length L in eqn. (4). At 2000 K the original spectrum has an average emissivity of 0.14. Increasing L from 1.4 to 2.8 m gives a value of 0.19. An average emissivity that is equal to the value of the spectrum after Ludwig *et al* (1973), i.e. 0.22, is found for L equal to 4.4 m. This path length is unrealistic, but it may be explained by differences in gas composition etc. in the emissivity computations. When L is increased, the emissivity rises much in the wavelength bands where ε is small, and little where ε is near 1. The effect is that almost clear windows in the spectrum become less transparent. Increasing the mean beam length L thus reduces the roof effect. When L is 4.4 m, the gain in the heat transfer is 3.9 %.

If the spectrum after the data by Ludwig *et al*, consisting of 372 wavelength bands, is used, the same trends are found as with the 15 bands spectrum. Increasing the (grey) roof emissivity from 0.4 to 0.95 enhances the heat flux by 3.6 % (no regenerator). This is nearly the same result as with the 15 bands spectrum with the same mean gas emissivity. With the regenerator we find a fuel saving of 2.4 %. The roof temperature increases somewhat less in these cases than with the 15 bands spectrum.

Using this gas spectrum, the experiment was carried out in which a rough spectral line structure was superposed on the spectrum. The effect of increasing the amplitude Δ can be seen in Figure 7. If Δ is raised from 0 to 0.4, the effect of the roof emissivity is seen to increase considerably. When the roof is completely black, the effects of the lines are almost leveled out (not completely because of some reflection at the glass surface). At lower refractory emissivities, however, the lines have a negative influence on the heat transfer to the load so that the savings are larger. The second experiment, in which the number of "lines" in a spectral interval is increased at constant amplitude Δ, showed that there is practically no influence of this number on the heat flux augmentation.

Apparently the details of the gas spectrum have influence on the heat transfer. For precise heat transfer calculations it is not enough to specify the average gas emissivity in a band. Of course, one has to be careful in assigning a specific representative value of Δ to a gas spectrum. One of the reasons is that in a gas the spectral lines have a form that is different from the simple block form that was used. However, the data from Ludwig *et al* show that in large parts of the spectrum the line width is much smaller than the line distance, so that the line effect discussed here can be expected to take place.

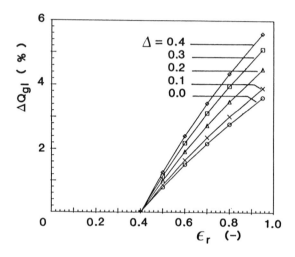

Fig.7. Effect of roof emissivity with the 372 bands spectrum for some values of Δ

Therefore, we conclude that the heat flux enhancement in the last situation is about 4 %, or, with the regenerator, the fuel saving is about 3 %.

Lastly, the roof effect was calculated with the Ludwig spectrum, increasing the emissivity from the measured spectrum to 0.95 (grey). The resulting fuel saving is about 2 % when the regenerator is taken into account. In a non-regenerative furnace we find a heat flux enhancement of about 3 %.

In Table 1 the results are summarized of these computations with the well-stirred furnace model.

8. HOW TO INCREASE THE REFRACTORY EMISSIVITY

In the previous section results have been given that show how much the furnace efficiency will increase when the refractory emissivity will be raised by a certain amount. The question of how this might be done, was still unanswered though.

Dark coatings of, for instance, silicon carbide or zircon - alumina - silica are used in steel industry (Fisher 1986). In glass industry they will lead to problems, because the quality of the glass is very sensitive

Table 1. Results of the computations with the well-stirred furnace model. Indicated are: the number of wavelength intervals of the gas spectrum, the mean beam length, the refractory emissivity in the initial situation, the heat transfer enhancement and the fuel saving (with regenerator). The final roof emissivity is always 0.95 (grey). In the last two rows, the amplitude Δ was increased from 0 to 0.4.

N_{bands}	L (m)	$\varepsilon_{roof,init}$	ΔQ (%)	$-\Delta\phi_{fuel}$ (%)
15	1.4	gray 0.4	7.9	5.3
15	2.8	gray 0.4	5.4	
15	4.4	gray 0.4	3.9	
15	1.4	spectral	5.4	
372	2.8	gray 0.4	3.6-5.6	2.4-3.8
372	2.8	spectral	2.5-4.4	1.6-3.0

to local additions of chemical reactants. The risk of a small quantity of coating material in the glass cannot be taken, because it would undo any saving at once. A different solution is therefore sought in creating holes or grooves in the roof. Especially V-grooves seem fruitful, because these can be applied without having any of the original flat surface left.

The effect on the apparent emissivity of infinitely long V-grooves has been calculated with a Monte-Carlo program, as shown in Figure 8. For different opening angles of the groove it is shown as a function of the emissivity of the flat surface. The results appear to be in good agreement with approximate analytical solutions (Hottel and Sarofim 1967). In sharp grooves radiation will be reflected more times so that these appear darker. Thus the problem is to create as sharp grooves as possible. Here it is not yet possible to state what minimum opening angle can be reached without knowledge of the thermal mechanical properties. This will be studied in future.

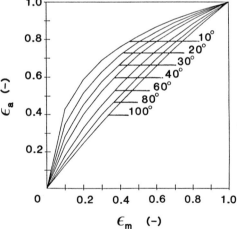

Fig.8. The effect of V-grooves on the apparent emissivity, as function of material emissivity and opening

9. CONCLUSIONS

Two different gas spectra have been used to predict the influence of the roof emissivity on heat transfer in natural gas fired high temperature furnaces, especially regenerative glass furnaces. An emissivity spectrum

of a typical glass furnace refractory was measured. From these computations and measurements the following conclusions can be drawn:
- At the temperature of the roof, the mean emissivity is about 0.4. The spectral characteristics of the refractory lead to a higher heat flux than a grey refractory.
- A high refractory emissivity is favourable. Increasing the refractory emissivity to 0.95 leads to a heat flux enhancement of about 3 % (without regenerator) or a fuel saving of 2 % (with regenerator). These results are obtained with the measured roof spectrum. A grey roof will lead to values that are about 1 % higher.
- The gas spectrum is the most important quantity that determines the importance of the roof emissivity. Only smoothed spectral emissivity data are not enough to predict the roof effect. Even detailed structures like the spectral lines in the gas spectrum can have influence on the outcome of heat transfer computations.
- In a small furnace the interaction of the spectral gas radiation and the roof will play a more important role than in a large furnace, because the transmitting "windows" in the gas spectrum are clearer.
- A regenerator compensates the spectral effects somewhat. A higher regenerator efficiency will have a stronger compensating effect.

LITERATURE

Alexander I, Gray W A, Hampartsoumian E, Taylor J M 1988 *Surface emissivities of furnace linings and their effect on heat transfer in an enclosure, Proc. 1st European Conf. on Industrial Furnaces and Boilers, Lisbon*

Carvalho M G, Durão D F G, Lockwood F C 1984 *Energy economics and management in industry, Proc. of the European Congress, Portugal*

Docherty P, Tucker R J 1986 *The influence of wall emissivity on the thermal performance of furnaces, Proc. Int. Gas Research Conf., Toronto* ed T L Cramer

Docherty P, Tucker R J 1986 *J. of the Inst. of Energy*, pp 35-7

Edwards D K 1960 *J.Opt.Soc.Amer.* 50 p.617-26

Elliston D G, Gray W A, Hibberd D F, Ho T-Y, Williams A 1987 *J. of the Inst. of Energy* pp 155-67

Fisher G 1986 *Ceram. Bull.* 65 2 pp 283-7

Goody R M 1964 *Atmospheric Radiation I, theoretical basis* (Oxford: Clarendon)

Gosman A D, Lockwood F C, Megahed I E A, Shah N G 1980 *The prediction of the flow, reaction and heat transfer in the combustion chamber of a glass furnace, Proc. AIAA 18th Aerospace Sciences Meeting, Pasadena*

Hottel H C, Sarofim A F 1967 *Radiative Transfer* (New York: McGraw-Hill)

Ludwig C B, Malkmus W, Reardon J E, Thomson J A L 1973 *Handbook of infrared radiation from combustion gases, NASA SP-3080* (Washington D.C.)

Oppenheim A K 1956 *Trans. ASME* 78 pp 725-35

Post L, Hoogendoorn C J 1988 *Proc. 1st European Conf. on Industrial Furnaces and Boilers, Lisbon* pp 1-8

Post L 1987 *Numerical methods in thermal problems* 5 1 eds Lewis R W, Morgan K, Habashi W G (Swansea: Pineridge) pp 884-95

Post L 1988 *Modelling of flow and combustion in a glass melting furnace* Ph.D. thesis, Delft University of Technology

Private communication 1988 *Report of test results* Department of Fuel and Energy, Wolfson Unit, University of Leeds

Soufiani A, Hartmann J, Taine J 1985 *J. Quant. Spectrosc. Radiat. Transfer* 33 3 pp 243-57

Tucker R J 1985 *Energy World* (nov.) pp 11-2

High emissivity coatings: are they energy efficient?

Dr N Pratten* D G Elliston° D F Hibberd° and D Thorpe+

* Energy Technology Support Unit, Harwell Laboratory, Oxfordshire, OX11 ORA
° British Steel Technical, Swinden Laboratories, Moorgate, Rotherham, S60 3AR
+ Unbrako Steel, Manor Road, Kiveton Park, Sheffield S31 8PB

ABSTRACT: This paper presents the background to a project which is currently being carried out at Unbrako Steel, Sheffield. The project, which is supported by the Energy Efficiency Office of the Department of Energy, aims to assess the effect of high emissivity coatings on the efficiency of furnaces. As the work is ongoing a more detailed paper will be presented on the day.

1. INTRODUCTION

The influence of high emissivity coatings on overall furnace performance has been a topic of recent debate. Initial laboratory studies and theoretical work failed to confirm manufacturers claims that these coatings enhanced energy efficiency. However, a recent investigation of the benefits of applying emissivity coatings to various types of furnaces (Elliston et al, 1987), has concluded that the surface emissivity of furnace walls will influence the net heat transfer to the stock. There are two main considerations which may alter the furnace efficiency.
(i) The surface may act to re-emit radiation at a different wavelength to that at which it is received. Because combustion gases absorb infrared radiation to varying degrees throughout the spectrum there are 'windows' at certain wavelengths where the gaseous medium is effectively transparent. If radiation is re-emitted within these windows then heat transfer to the stock will be increased because absorption losses will be smaller. Thus the furnace efficiency will increase.
(ii) The absorption characteristics of the coating may be such that more heat is conducted through the furnace walls as a result of applying the coating. This will result in a fall in efficiency.
It has been suggested that the relative magnitudes of these competing effects will determine whether the application of high emissivity coatings is beneficial or detrimental to furnace efficiency.

Ellison at al suggested that the most promising application of high emissivity coatings would be to a furnace indirectly heated by combustion gases, such as a controlled atmosphere heat treatment furnace. The Energy Efficiency Office through its EEDS operated by ETSU, have therefore supported a programme of work to carry out independent monitoring trials of the performance of coatings on a bell type annealing furnace. This will be the first controlled study on a full scale plant of this type and will demonstrate unambiguously, the potential energy benefits.

2. PROGRAMME

The programme of work in this study is being performed at Unbrako Steel, Sheffield. Unbrako Steel employ Lee Wilson batch heat treatment furnaces with inert atmospheres for annealing steel wire at 760°C, (Fig.1). It is intended that a single furnace unit (consisting of a base, an inner cover and an outer cover housing 20 gas fired radiant tubes), will be coated in stages and monitored to assess any changes in furnace performance. The coating will be the ET-4 high emissivity coating, which is a zirconium based product manufactured by the Ceramic Refractory Corporation. The coating suppliers have estimated that an energy saving in the range of 15% to 35% may be achieved.
Monitoring of stock temperature, flue gas composition and temperature, in addition to gas consumption is being carried out by British Steel Technical, Swinden Laboratories. The furnace efficiency will be calculated from the heat balances, to determine the independent effects of coating radiant tubes and inner furnace cover. In addition to energy benefits, the protective quality of the coating to the cover and radiant tubes will be assessed. This is an important property of the coating which has been claimed to provide a sixfold increase in radiant tube lifetime. All benefits considered, it is thought that a high emissivity coating applied successfully to the bell type furnace could have a payback period of less than 3 years.
This ongoing programme of work should demonstrate unequivocally, the effect of high emissivity coatings on the performance of the bell type furnace and results obtained to date will be discussed.

3. REFERENCES

Elliston D G, Gray W J, Hibberd D F, Ho T-Y, and Williams A, 1987. The effect of surface emissivity on furnace performance. Journal of the Institute of Energy, 155-167.

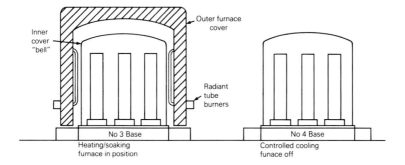

Figure 1.   Bell Type Annealing Furnace

Development and applications of ceramic electrochemical reactors

B.C.H. Steele

Centre for Technical Ceramics,   Department of Materials,
Imperial College of Science, Technology & Medicine,   London   SW7 2BP.

ABSTRACT:
The principal types of fuel cells under development are briefly reviewed, and the possible configurations for solid oxide fuel cells (SOFC) are outlined.   The current status of SOFC units is then summarised, and particular attention is given to the processing, microstructure, performance relationships of the ceramic components incorporated in SOFC stacks, which provide one example of ceramic electrochemical reactor technology.   Other examples, including the high temperature electrolysis of steam, separation of oxygen from air, and the partial oxidation of methane to form ethane and ethylene, are also briefly discussed.

1.   INTRODUCTION

The main features of the five principal fuel cells under development are summarised in Figure 1, and further details are provided in the Handbook of Fuel Cells by Appelby and Foulkes (1989).   The two high temperature systems, **molten carbonate fuel cell (MCFC)** operating at 650°C, and **solid oxide fuel cell (SOFC)** operating at 900–1000°C, can in principle convert hydrocarbon fuels directly into electricity at high efficiencies (55–60%) with low NO_x emission.   Moreover, the associated high temperature, high grade heat available from MCFC and SOFC systems, favour their use in a variety of industrial co–generation situations.   An additional useful feature of the SOFC system is the ability to be operated in different modes, thus spreading the development risks over different market sectors.   These and other factors are making ceramic SOFC systems attractive for future energy technology investment.

A variety of configurations are now being considered for the solid oxide fuel cell (SOFC) stack.   These include tubular, monolithic and planar types, and within each type different manufacturing routes can often be chosen to fabricate the completed assembly.   The interacting influences of design, materials selection, processing route and performance are not always appreciated by the design engineer, and more interaction is required between the various experts to avoid unnecessary expense and possible inability of the component to meet design requirements.   The various stages in the design of a metallic component have often been discussed (Charles, 1989) and a variety of databases and expert systems are available to optimise materials selection.   However, for the materials incorporated in SOFC stacks very little high temperature data are available to assist in the design of the brittle ceramic components, and at present development is largely empirical.   In a recent paper Gauckler and Kummer (1989) have emphasised that ceramic materials exhibit slow crack growth even under moderate stress levels and this phenomenon can produce catastrophic failure after a certain time.   Crack growth velocities are related to stress intensity factor (K_{IC}) of a material.   Therefore, the time for which the material can withstand a certain stress level under static load can in principle be estimated.   Finite element stress analysis, together with statistical data of rupture strength and slow crack growth can thus predict the probability of failure, depending on location and time.

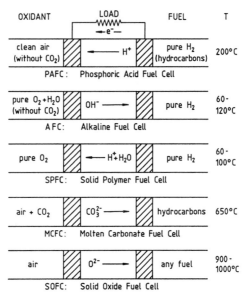

Figure 1: The main features of the five principal fuel cells under development.

A further feature of the mechanical properties of ceramics is the strong R curve behaviour (Figure 2) in which the fracture toughness exhibits a strong dependence upon crack length.   It is therefore necessary to question the wisdom of fabricating very thin ($<10\mu m$) SOFC stack components.

Figure 2:
Schematic diagram showing
typical R–curve behaviour
for a zirconia ceramic.

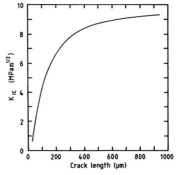

These and other relevant properties of SOFC stack components will be discussed for the following materials that are likely to be incorporated into the first generation of SOFC units, namely:

Electrolyte :	$Zr_{1-x}Y_xO_{2-x/2}$
Cathode :	$La_{1-x}Sr_xMnO_{3-\delta}$, $La_{1-x}Ca_xMnO_{3-\delta}$
Anode :	$Ni - ZrO_2$ (approx. 50% by vol)
Interconnect : (bi–polar plate)	$La_{1-x}Sr_xCoO_{3-\delta}$, $LaCr_{1-x}Mg_xO_{3-\delta}$

The influence of the cell design on the fabrication route and subsequent performance of these materials will be summarised for the three principal configurations, and some recent results obtained at Imperial College for components to be incorporated into the planar configurations will be presented.

2. TUBULAR CONFIGURATION

In recent years, development of the tubular configuration has been dominated by Westinghouse who have made impressive progress in the manufacture and operation of 3kW systems, and the design and fabrication of 25kW units is now underway. The principal features of the Westinghouse supported tubular design (Figure 3) are well known, and the configuration adopted avoids the necessity of having high temperature seals. Moreover, for in–situ reforming operation, the fuel / spent fuel mixture can be distributed along the length of the tubular cell to reduce local low cell temperatures arising from the endothermic reforming reaction, thus avoiding severe thermal stresses within the ceramic components. The principal queries associated with the Westinghouse design are concerned with scale–up of the fabrication route. Can the porous supporting tube be increased in length to 1m, as proposed, without sacrificing reliability and low cost? More importantly, can the relatively slow CVD/EVD process for depositing the thin layers of electrode and electrolyte materials be scaled–up for mass production, or will it be necessary to switch to other fabrication methods such as plasma spraying technique used by the Electrotechnical Laboratory in Japan?

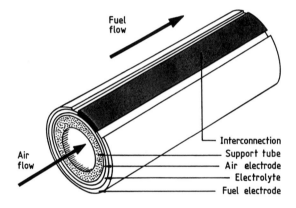

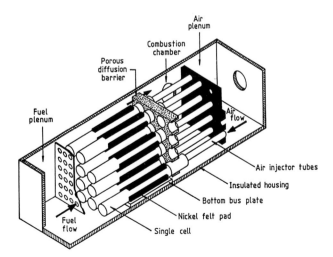

Figure 3: Westinghouse tubular SOFC: single tube arrangement (top) and tube bundle arrangement (bottom).

The tubular configuration adopted by Westinghouse emphasises how the design influences the processing route as mehods easily amenable to mass production, such as screen printing, tape casting, etc. cannot be used on curved surfaces. Moreover, selection of the cathode composition which is believed to be $La_{0.85}Sr_{0.15}MnO_{3-\delta}$ may be dictated by the CVD processing route employed. It is known that the electronic conductivity of either $La_{0.5}Sr_{0.5}MnO_3$ or $La_{0.5}Ca_{0.5}Mn_{3-\delta}$ is higher than the composition $La_{0.85}Sr_{0.15}MnO_3$. Furthermore, it has been reported by Takeda (1987) and Steele et al (1988) that the cathodic polarisation produced during oxygen reduction is also lower for the high doped material, and so it appears that the Westinghouse design incorporates a composition with inferior properties. Problems arising from thermal expansion mismatch cannot explain the decision to use compositions with lower dopant concentrations, and so the explanation is probably associated with the fabrication route used to manufacture the ceramic stack components. It is possible that problems with CVD technique limit the strontium oxide doping concentrations to relatively low levels due to the gaseous chemical reactions involved or due to interfacial reactions involving strontium oxide at the high temperatures involved in the CVD deposition process.

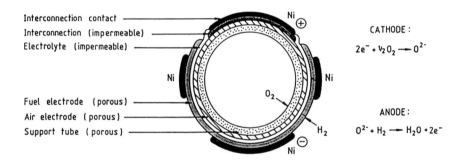

Interconnection contact

Interconnection (impermeable)

Electrolyte (impermeable)

Ni

Ni

Fuel electrode (porous)

Air electrode (porous)

Support tube (porous)

O_2

H_2

Ni

CATHODE :

$$2e^- + \tfrac{1}{2}O_2 \longrightarrow O^{2-}$$

ANODE :

$$O^{2-} + H_2 \longrightarrow H_2O + 2e^-$$

Figure 4a: Section through Westinghouse tubular SOFC.

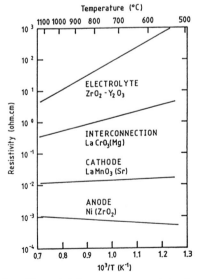

Figure 4b: Specific resistivity–temperature behaviour of SOFC.

The manner in which design can influence performance is illustrated further by continuing examination of the cathode component. In the Westinghouse configuration the cathode current collection path is relatively long (Figure 4a), and so this component produces the largest ohmic polarisation loss (Table 1), even though the specific resistivity of $La_{0.85}Sr_{0.15}MnO_{3-\delta}$ is significantly lower (Figure 4b) than that of the electrolyte or interconnect material.

TABLE 1

OHMIC POLARISATION LOSSES

(Assuming uniformly distributed current through the electrolyte)

Material	Specific resistivity at 1000°C (ohm.cm)	Material thickness (mm)	Contribution to cell resistance (%)
Cathode ($LaMnO_3$–Sr)	0.013	0.07	65
Anode (Ni+Zr)	0.001	0.1	25
Electrolyte	10.0	0.04	9
Interconnection ($LaCrO_3$–Mg)	0.5	0.04	1

OD of support tube = 13mm;
length of cord = 35mm for electrolyte, 6mm for interconnection

The self–supported tubular design (Figure 5) favoured by Dornier may be more amenable to mass production. This design involves the separate fabrication of small, hollow ceramic cylinders of the electrolyte, interconnect and insulator, which are subsequently sintered together to produce an impermeable tube. The cathode and anode layers are then applied as a slurry after appropriate masking of the interconnect and insulating regions. This design has so far been successfully assembled into 2kW units.

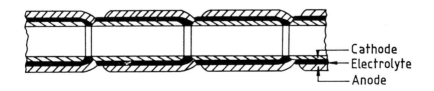

Figure 5: Schematic diagram of an unsupported multicell design.

3.   MONOLITHIC   CONFIGURATION

The Argonne National Laboratory have pioneered the concept of the monolithic solid oxide fuel cell (Figure 6), and demonstrated the feasibility of this concept by fabricating and testing small (5 x 5cm) four cell stacks.   The drying and co-firing of thin layers of the impermeable electrolyte and bi-polar plate materials together with the porous electrodes, represents a major technological challenge, as the drying and sintering shrinkages must be closely matched together with the thermal expansion coefficients. Scale-up of the fabrication process is being investigated by Allied Signal in collaboration with Combustion Engineering.   A major problem to be overcome by the process engineers is to achieve a uniform input of heat into what is essentially an insulating structure composed of materials with different thermal conductivities.   This problem has been recognised and a patent has been published by Riley (1989), which suggests that microwave sintering could be used to fabricate the monolithic structure.   However, the different electrical susceptibilities exhibited by various components could introduce formidable difficulties, and it will certainly be interesting to observe developments in the scale-up of the monolithic configurations.

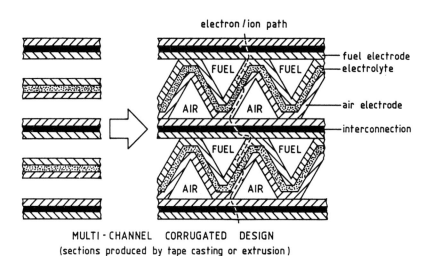

MULTI - CHANNEL   CORRUGATED   DESIGN
(sections produced by tape casting or extrusion)

Figure 6:   Argonne monolythic concept.

4.   PLANAR   CONFIGURATIONS

4.1   General Comments

In principle, the planar configuration (Figure 7), which is usually adopted for other fuel cells operating at lower temperatures, offers many advantages over the other SOFC configurations.   Compared to the Westinghouse tubular configuration, the planar structure should develop higher power densities, and deposition of the porous electrodes can be accomplished by mass production methods such as tape casting or screen printing.   A further advantage is that the various components such as the anode | electrolyte | cathode structure and the bi-polar plate can be independently checked for faults prior to assembly.   Quality control can thus be exercised on the individual components rather than on the processing procedures, which will be imperative for the Westinghouse tubular and the Argonne monolithic configurations.   It may also be possible to specify alternative materials for the bi-polar plate, and Z Tek, Siemens and National Chemical Laboratory of Japan are already investigating the behaviour of metallic bi-polar plates.

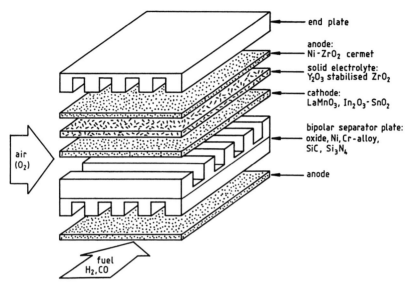

end plate

anode:
Ni-ZrO$_2$ cermet

solid electrolyte:
Y$_2$O$_3$ stabilised ZrO$_2$

cathode:
LaMnO$_3$, In$_2$O$_3$-SnO$_2$

bipolar separator plate:
oxide, Ni, Cr-alloy,
SiC, Si$_3$N$_4$

anode

air
(O$_2$)

fuel
H$_2$, CO

Figure 7:   Cross-flow planar SOFC design.

However, these advantages may be offset by a number of disadvantages which include the necessity for high temperature seals, and the fact that the components will be more highly stressed.   Mechanical stresses will be developed by the seals, and thermal stresses by the temperature differences caused by local cooling due to the endothermic in-situ reforming reaction and possible localised Joule heating.   It will be necessary, therefore, to ensure that the principal structure components, namely the solid electrolyte and bi-polar plate, have appropriate high temperature mechanical properties.   Consequently, this is one of the principal areas of investigation at Imperial College, the other being the development of alternative anode materials, which was described at Hakone (Steele, 1989).

4.2   Mechanical Behaviour of SOFC Components

Little information has been published relating to the mechanical properties of SOFC components.   Rossing (1983) has suggested that room temperature fracture strengths of 43MPa for the porous calcia-stabilised support tube used in the Westinghouse tubular design would be sufficient for fuel cell fabrication and operation.   This assertion assumes that the high temperature strength at 1000°C would be around 75% of the room temperature value.   Bosak et al (1988) describe procedures for improving the fracture toughness (K_{IC}) of the zirconia electrolyte used in the Argonne monolithic configuration and, by using partially stabilised zirconia electrolytes, room temperature K_{IC} values in the range 3-4MPam$^{\frac{1}{2}}$ were reported.   However, these compositions exhibited lower ionic conductivities than pure cubic ZrO$_2$ electrolyte materials.

At Imperial College we specified the following preliminary properties for the electrolyte:

Specific resistivity:	< 10ohm.cm at 900°C
Fracture strength:	> 400MPa at room temperature
Sheet dimensions:	5cm x 5cm x 200μm

In collaboration with U.K. industrial laboratories we have examined both tetragonal and cubic zirconia electrolyte sheets, fabricated to the above dimensions using various commercially available powders and two processing routes, namely tape-casting and extrusion.   The results are summarised in Table 2.

TABLE   2

Summary of conductivity and resistivity data for ZrO_2-Y_2O_3 electrolyte sheets

| Powder Source | Comp m% Y_2O_3 | Conductivity | | | | Resistivity (900°C) | | |
| | | Bulk (B) | | Grain boundary (GB) | | | | |
		$\log\sigma_0$ S.cm^{-1}	E_a eV	$\log\sigma_0$ S.cm^{-1}	E_a eV	α_B Ωcm	α_{GB} Ωcm	α_T Ωcm
Mandoval (DK)	3	5.41	0.83	5.74	0.98	16	44	60(E)
Toya Soda	3	5.63	0.86	5.94	1.05	16	4	20(E)
Daiichi Kigenso	3	5.55	0.87	6.44	1.01	20	11	31(E)
Toya Soda	8	6.37	0.96	7.23	1.11	6	4	10(E)
Toya Soda	8	6.90	1.01	7.14	1.02	4	3	7(T)
Toya Soda	6	5.73	0.89	6.47	1.06	17	12	29(T)
Z. Tech	4–5	6.17	0.91	7.03	1.02	10	1	11(T)

E = Extruded   T = Tape–cast

Examination of this table indicates that the specific resistivity of the cubic (8 mole% Y_2O_3) electrolyte can achieve the required value (<10ohm.cm at 900°C) when fabricated by either process route.   However, only the electrolytes prepared by the extrusion route were able to achieve the required mechanical properties.

Further evidence of the superior mechanical properties generated by the extrusion route is evident from Figure 8, which shows a Weibull plot for cubic zirconia electrolytes fabricated by the two process routes.   The range of fracture strenghts recorded for the extruded material is relatively small, as indicated by the high Weibull moduli (m=16), whereas the number and distribution of critical flaws in the tape–cast material is much larger (lower Weibull modulus), indicating that this material is relatively unreliable for design calculations.

Figure 8: Weibull plots for ZrO_2-Y_2O_3 (8 mole %) electrolyte sheets prepared by: (a) tape casting, (b) extrusion.

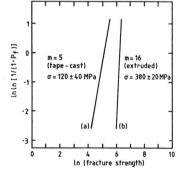

Work is in progress at Imperial College to measure the high temperature mechanical properties of the extruded material, and preliminary results indicate that the fracture strength at 900°C is about 60% of the room temperature value.   Slow crack growth under cyclic fatigue and cell operating conditions is also being monitored.

Single cells have been assembled using the high strength extruded electrolyte sheets (5cm x 5cm x 200μm). 100μm thick porous electrodes of $La_{0.5}Sr_{0.5}MnO_3$ (cathode) and $Ni-ZrO_2$ (anode) were deposited on the electrolyte sheet by tape–casting routes. Subsequent firing at 1200°C produced excellent adherent structures having about 50% porosity with a medium pore size around 3μm. The coated electrolyte sheets were also sealed into a Macor holder, as shown in Figure 9, and the I–V characteristics measured. At 0.7V the current density was 120mA.cm^{-2} at 900°C for 80% conversion of the hydrogen fuel. Assuming purely ohmic I–V characteristics, the derived internal cell resistance was 5.8ohm.cm^{-2}, which compared to a value of 0.4ohm.cm^{-2} calculated from the specific resistivity values of the electrode and electrolyte materials. Studies are in progress to identify the factors responsible for the higher than expected values for the internal resistance.

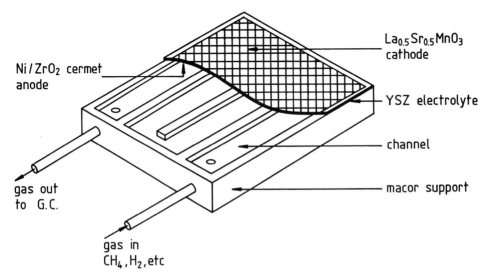

Figure 9: Single cell arrangement incorporating 5cm x 5cm x 200μm $ZrO_2-Y_2O_3$ (8 mole %) electrolyte plate.

Investigations have also commenced into the mechanical properties of the bi–polar plate material, $LaCr_{0.8}Mg_{0.2}O_{3-\delta}$. Problems were initially encountered using powder purchased from the U.S.A., and so we have developed our own powders at Imperial College. Using these powders, dense (>97%) bars of bi–polar material could be obtained by sintering in air at 1500°C. Preliminary experiments indicate fracture strengths around 300MPa at room temperature and the high temperature mechanical behaviour is now under examination.

5. OPERATIONAL MODES OF CERAMIC ELECTROCHEMICAL REACTOR

A major advantage of the SOFC system is its ability to operate in a variety of modes, as indicated in Table 3. It can be used as a high temperature electrolyser to produce hydrogen, and Dornier in Germany have constructed a 2kW unit for evaluation of this process. Alternatively, the ceramic reactor can be used to directly electrochemically oxidise natural gas to water and carbon dioxide (fuel cell mode) or to partially oxidise methane to produce useful chemical feedstocks (Steele at al, 1988) such as ethane and ethylene (chemical production mode). Other possibilities are summarised in Table 3, and it is anticipated that proton conducting ceramic electrolytes (Iwahara, 1988) may also become economically attractive to provide further interesting possibilities for ceramic electrochemical reactors.

TABLE 3

Ceramic   Electrochemical   Reactors

ELECTRODE 1	CERAMIC ELECTROLYTE	ELECTRODE 2
$\frac{1}{2}O + 2e \rightarrow O^{2-}$	$---\ O^{2-}\ \longrightarrow$	$O^{2-} + H_2 \rightarrow H_2O + 2e$
$O^{2-} \rightarrow \frac{1}{2}O_2 + 2e$	$\longleftarrow--\ O^{2-}\ ---$	$H_2O + 2e \rightarrow H_2 + O^{2-}$
$\frac{1}{2}O_2 + 2e \rightarrow O^{2-}$	$---\ O^{2-}\ \longrightarrow$	$O^{2-} + \frac{1}{4}CH_4 \rightarrow \frac{1}{4}CO_2 + \frac{1}{2}H_2O + 2e$
$\frac{1}{2}O_2 + 2e \rightarrow O^{2-}$	$---\ O^{2-}\ \longrightarrow$	$O^{2-} + 2CH_4 \rightarrow C_2H_6 + H_2O + 2e$
$\frac{1}{2}O_2(N_2) + 2e \rightarrow O^{2-}$	$---\ O^{2-}\ \longrightarrow$	$O^{2-} \rightarrow \frac{1}{2}O_2 + 2e$
$H^+ + e \rightarrow \frac{1}{2}H_2$	$\longleftarrow--\ H^+\ ---$	$\frac{1}{2}C_2H_6 \rightarrow \frac{1}{2}C_2H_4 + H^+ + e$

6.   CONCLUSIONS

The interacting influences of design, materials selection, processing route and performance have been discussed for the three principal SOFC configurations.   The importance of the processing route has been illustrated by presenting mechanical property data for ceramic zirconia electrolyte sheets prepared by tape–casting and extrusion routes.

Finally, it should be emphasised that the combination of high electricity / fuel conversion efficiencies (55–60%), low enviromental pollution, quiet operation and modular construction, all indicate that fuel cell systems are a technology whose time has finally arrived after 150 years!

7.   ACKNOWLEDGEMENTS

The author wishes to acknowledge the work of the following members of the Centre for Technical Ceramics: Dr P.H. Middleton, Dr A.C. Leach, Mr R. Rudkin, whose results were quoted in this paper.   I also wish to express thanks to the EEC for the contract EM3E/0167/E which provided partial financial support for the work reported here.

8.   REFERENCES

Appleby A J and Foulkes F R   1989   *Fuel Cell Handbook*   (Van Nostrand Reinhold)

Bosak A L, Singh J P, Dess D W and McPheeters C C   1988   *Abs. of 1988 Fuel Cell Seminar, Long Beach, California, October 23–26th 1988*   (Washington DC: Courtesy Associates Inc.) p.145

Charles J A   1989   *Mat. Sci. and Tech.* **5** 509

Gauckler L and Kummer E   1989   *Proc. of Workshop on Mathematical Modelling of Natural Gas Fuelled Solid Oxide Fuel Cells and Systems, Int. Energy Agency, Charmey, Switzerland, July 2–6th 1989* (Berne: ENET, c/o OFEN) p.143

Iwahara I   1988   *Solid State Ionics* **28–30** 573

Riley B   1989   *Process of Forming Conductive oxide Layers in Solid Oxide Fuel Cells* U.S. Patent 4,799,936 (January 24th 1989)

Rossing B R   1983   *Proc. of Conf. on High Temperature Solid Oxide Electrolytes, Brookhaven, August 16–17th 1983* (Brookhaven Nat. Lab. Publ. BNL–51728) p.43

Steele B C H, Kelly I, Middleton H, and Rudkin R   1988 *Solid State Ionics* **28–30** 1547

Steele B C H   1989   to be published in *Proc. of 7th Int. Conf. on Solid State Ionics, Hakone, Japan November 5–11th 1989*

Takeda Y, Kanno R, Noda M, Tomida Y and Yamamota O   1987   *J. Electrochem. Soc.* **134** 2656

Ceramic tape casting for solid oxide fuel cell (SOFC) electrolyte production

S V Phillips, A K Datta

GEC ALSTHOM Engineering Research Centre,
P O Box 30, Lichfield Road, Stafford. ST17 4LN

1. INTRODUCTION

The solid oxide fuel cell (SOFC) is one of the most recent fuel cells devised for the direct conversion of (gas or vapour phase) fuels to electricity. The main attractions are that such cells do not utilise a thermodynamic cycle and thus are not limited by Carnot efficiency,and are capable of operation at high temperatures,typically 800 to 1000°C.This allows incorporation into gas turbine units and primary heat sources for production of combined heat and power. Estimates of electrical efficiency of such plants are realistically 50% (at 85% fuel conversion) compared with ~30% for conventional systems. There are problems associated with high temperature operation connected with materials stability and compatibility, temperature control and fuel/air handling. These problems are being tackled in the USA, where Westinghouse and Argonne National Laboratories are amongst the leaders in SOFC development; in Japan SOFC development is active under the Moonlight Plan although it is less advanced than phosphoric acid or molten carbonate cells. European effort is being co-ordinated through CEC programmes involving both industry and universities and the next phase of this work under the 'Joule' initiative is directed towards a three kW SOFC stack. The work at the GEC Alsthom Engineering Research Centre, Stafford was part funded by the CEC, with most of the effort being directed to the practical problem of producing first, large area electrolyte plates, and then channelled electrolyte stacks which could form the base structure for a SOFC. This paper describes work on tape casting related to this problem, considered in the following sections:

1 - requirements for SOFC electrolytes

2 - description of tape casting

3 - process details   a) casting
                       b) firing

   - results and materials properties

   - SOFC construction and future prospects

2. THE TAPE CASTING PROCESS

The process requires a slip or suspension of the ceramic particles dispersed in a liquid. This consists of dissolved organic binders and plasticizers taken up in a solvent system. This system is specifically designed for the ceramic powder to be cast, to enable the particles to remain in suspension during preparation and casting.

The slip is spread or cast in a controlled manner on to a flat surface and the solvents allowed to evaporate. The remaining binder and adhering particles form a handleable sheet, like a paper or leather product, known as 'green' (ie unfired) tape. One of the first industries to utilise tape casting was the capacitor industry, where the method was ideally suited to producing thin sheets of high permittivity ceramic, required for the quality production of discrete capacitors. These tapes or sheets could be as thin as 25μm. In microelectronics substrates and packaging, tapes 0.25 to about 2mm thick made of alumina, beryllia and other dielectric materials are cast, and passive components delineated on the substrates by thick film screen printing. With the move to higher operating frequencies low permittivity substrates are required and glass-ceramics with this property have been tape cast at Stafford and elsewhere. The requirements of the SOFC have resulted in tape casting of doped zirconia powder suspensions and the processing of individual sheets into the cross flow configuration of a multiple stack. It is apparent that tape casting was a versatile production method, because it allowed post-casting shaping and joining processes to be carried out on the 'green' tape, often without further additives to ensure bonding of the several parts on firing the assembled article.

Preparation of the suspension, casting and firing of yttria-stabilised zirconia are described in the following sections.

The suspending medium consists of several components:

1) Binder

This is used to form an adhering film round the powder particles and to impart strength and handleability to the green tape, and to decompose slowly on heating so as not to disrupt the tape in the pre-sintering stage. Common binders are PVB (polyvinyl butyral) and acrylic polymers, with 2% to 6% by weight of binder being added to the inorganic powder.

2) Dispersants

Maintaining particles of the ceramic oxide in suspension, and avoiding flocculation which would accelerate settling is of prime importance, and is achieved by modifying the electrical double layer surrounding each particle. The highest efficiency dispersants are specific to particular powders and several researchers have examined a wide range (Ref 1); However, a natural product, Menhaden herring oil as a mixture of fatty acids, is effective for a variety of oxide powders.

3) Plasticisers

To reduce the viscosity of the binder, a plasticiser is added, typically polyethylene glycol (PEG - molecular weight 1000 to 1500) in quantities equal to that of the binder.

4) Solvents

The solvent dissolves the binder and other components, and must evaporate slowly after the tape is cast. Usually mixtures of alcohols, aromatic hydrocarbons and chlorinated solvents are used to achieve a compromise between an acceptably high evaporation rate and avoidance of surface 'skin' formation which traps gas bubbles.

5) Aqueous systems

The above systems are all based on organic compounds. Water soluble binders, such as some acrylics and celluloses have obvious advantages in terms of safety and lower cost, but there are disadvantages typically increased drying times and temperatures. The stability of the green tape during firing has been found to be improved for certain doped zirconia powders by the use of water based systems.

6) Casting systems

The system used at ERC for zirconia tape casting is illustrated in Fig 1. The hopper fed from a reservoir of slip has a gate or doctor blade fixed at a predetermined height above the casting 'belt'. This has consisted of a melinex strip, wound onto a motor driven drum which allows about 3 metres to be cast. By stripping the tape longer lengths are produced; for laboratory scale 1 to 3 metres is sufficient. It is possible to use other casting substrate materials, and float glass is very satisfactory.

Drying of the cast tape is accelerated by air flow; in the case of water based systems drying may take 8 to 12 hours.

3. TAPE CASTING OF YTTRIA-DOPED ZIRCONIA

The general principles adopted in tape casting have been outlined in Section 2. Some details applicable to the production of Y_2O_3/ZrO_2 are as follows:

3.1 Supply of powder

For fuel cell purposes Y_2O_3-doped ZrO_2 powders were bought from commercial suppliers. Of the several available, the following were used for tape casting:

- Tosoh TZ-8Y fully stabilised ZrO_2 containing 8 mol % Y_2O_3. Average particle size 0.3μm, specific surface area 20 m^2/g.

- Z-Tech SY.6 mol% Y_2O_3. Average particle size 0.25μm.

The first consideration determining the doping level was the conductivity of the fired ceramic. The conductivity at the cell operating temperature is discussed later, but it is clear that it must be maximised for reduced loss in the operating fuel cell. Secondly, the Y_2O_3 level was maintained at or above the level to ensure all the zirconia is in the cubic form. It appears this minimum level of dopant (in this case Y_2O_3) is close to that required for maximum conductivity.

3.2 Tape casting details

The casting gate was set at 1.5mm to yield green tape of ~0.8mm thickness. It was found that casting on to glass plates (float glass) gave more uniform separation of the tape than from a plastic film base (silicone coated Melinex). In addition the 'downward' surface of the tape, contacting the glass was smoother than the 'top' or free surface.

After drying,the tape was stored flat in plastic sleeves, typically in 0.5 x 0.2m lengths. For ribbed plates, these were shaped using a square-wave section plate, grooved 1mm wide x 1mm deep, at 4mm pitch. The 30% shrinkage on firing reduced these dimensions proportionally. The dimensions chosen were somewhat arbitrary at this stage and it is considered the number of ribs per plate forming a cell could be reduced while maintaining a stable stack. Theoretical studies on gas flow in the channels will clearly assist in establishing dimensions.

3.3 Firing

The firing schedule consists generally of two parts

- low temperature binder burnout

- high temperature sintering stage

The burnout stage was investigated by TGA of the green tape using a DuPont Thermal Analyser. This identified critical temperature regions where $\underline{dT}$ was kept low in the schedule.
                    dt

The sintering temperature region was investigated from 1350°C to 1550°C, the target being to eliminate bulk porosity. A suitable schedule was arrived at by experimentation,the final temperature being 1520°C. Other factors investigated were the nature of the substrate, which influenced tape adhesion during firing and the incidence of bowing of single plates.

Stacks of ribbed plates shown in the green state in Fig 2 were fired to give cross flow configurations varying in size from ~25 x 25 x 20mm to 50 x 50 x 45, as shown in Fig 3. The larger stack has a volume in excess of 100cm^3 with an area to volume ratio of some 1300 m^2/m^3. This high ratio, about five times that of a tubular SOFC, allows for a high power density of a cross flow stack; with obvious advantages. Equally, temperature monitoring and control becomes more important in such a structure, and exploiting the versatility of tape cast construction has enabled a cross flow structure with gas flow temperature control to be devised at ERC (Ref 2). The relative importance of the electrical, mechanical and thermal properties of the doped ceramic depends to some extent on the construction methods adopted for the fuel cell structure. With the monolithic cross flow electrolyte structure the important properties are:

- high electrical conductivity

- adequate mechanical strength to provide a robust structure

- compatibility with electrode materials

- ability to be fabricated in the sizes required, and cheaply

- lifetime of the stack complex

i) Electrical conductivity

The conductivity of pure ZrO_2 is mainly electronic, but with the addition of dopants such as CaO or Y_2O_3 to stabilize cubic ZrO_2, oxygen vacancies are generated. At a pO_2 of 10^{-7} atmospheres, log σ is -2.5 for pure ZrO_2, compared with a log σ of -1.0 for ZrO_2 doped with 15m% CaO (Ref 3) the conduction being due entirely to oxygen vacancies. High conductivity is also attained with 8m% Y_2O_3 doping. Measurements provided by Imperial College on fired tape cast material made at GEC ALSTHOM, Stafford (3m% and 8m% Y_2O_3/ZrO_2) showed conductivities of c.10^{-2} Scm^{-1} and 0.2Scm^{-1} respectively at 900°C and extrapolation of the latter to 1000°C gave about 0.5 Scm^{-1}. Work by the Instituto de Ceramica y Vidrio in Madrid on the effect of mixed stabilizing oxides on ZrO_2 conductivity has resulted in $\sigma_{1000°C}$ of about 1 Scm^{-1}, believed to be the highest so far recorded.

ii) Mechanical Strength

The mechanical strength of the cast tape, measured as the rupture modulus in 3 point bending of 'bars' cut from tape was 135Nmm^{-2}, SD 28Nmm^{-2} compared with 300 Nmm^{-2} for bars of the same material, 8m% Y_2O_3-ZrO_2. Whereas the mechanical strength of the ceramic needs to be a maximum, the overall design of the stack including manifolds must be aimed at minimising thermal shock stresses in order to obtain the best performance.

iii) Electrodes

Trials of Stack assembly have been done with bare ZrO_2 tape, ie with no electrodes. Working cells require anode and cathode in contact with the electrolyte and in some configurations a bipolar plate to allow series connection of adjacent cells. The oxygen electrode (cathode) is La-Mn-Sr-O ceramic available as a powder, conductivity typically 100 to 300 Scm^{-1} at 1000°C. This can be applied, by printing techniques suitable for production, to one face of the fired zirconia tape.

There are several candidate materials for the fuel electrode (anode), the most promising being a nickel-zirconia cermets. This may be applied by screen printing onto the zirconia tape. Typical conductivity values for ~40 vol% Ni cermets are ~ 7-10 S cm^{-1} (Ref 4) although recent work has increased this value.

iv) Bipolar plate

A possible material is La-Cr-O with high electron conductivity at 1000°C, which again could be applied by printing, or used as a cast tape.   Refractory metal alloys are also under consideration.

4. CONCLUSIONS AND OUTLINE OF FURTHER DEVELOPMENTS

Tape casting has been shown to be very suitable for the production of doped zirconia plates which form the electrolyte, and the structural foundation of a practicable Solid Oxide Fuel Cell. Green tapes, both plane and ribbed have been cast on a semi-continuous basis, and there appears no reason why the process so far developed at ERC, Stafford should not be extended to large scale production.   Considerable engineering expertise will need to be applied to realise the large area of plates required. A reasonable estimate of the power density for an SOFC is 250kW/m^3, so a 1 MW stack would need some 5200 square metres of zirconia tape. For a tape width of, say, 25-26 cm this corresponds to some 20 km of tape; fortunately these quantities have been handled by ceramic capacitor manufacturers so the technology is well developed. In addition to the electrolyte tape the electrodes and bipolar plate are to be produced in proportionate amounts and here the technology is less advanced. With the effort and funding currently allocated to SOFC in Europe and in the rest of the world these problems will be overcome.

5. ACKNOWLEDGEMENTS

The authors express thanks to GEC ALSTHOM Ltd for permission to publish this paper. We also acknowledge the support of the Commission of the European Communities for work on Solid Oxide Fuel Cell development, Project EN 3E/0167, and the assistance of other project partners.

6. <u>REFERENCES</u>

1. K. Miseka, R. Cannon, 'Forming of Ceramics' Vol. 9, 164 American Ceramic Society Inc., Columbus U.S.A.

2. U K Patent Application No. 8812641.2

3. A.M. Anthony 'Science and Technology of Zirconia - Vol 3. American Ceramic Society Inc. Columbus USA.

4. R. Dees J. Electrochem Soc. 134 2141, 1987

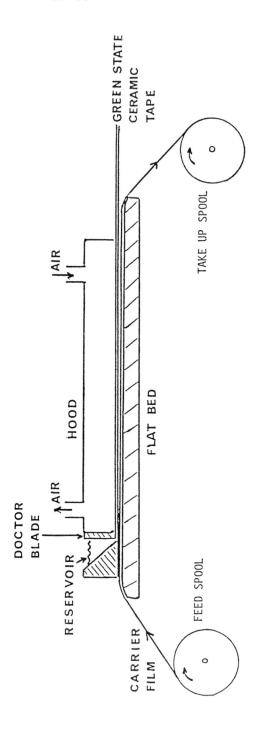

Figure 1 Diagram of tape casting equipment.

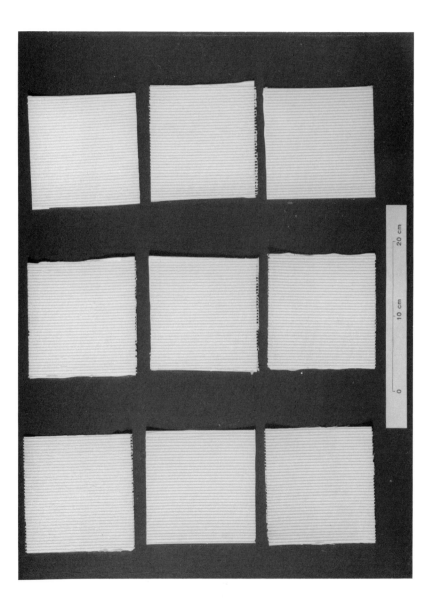

Figure 2 $Y_2O_3 \cdot ZrO_2$ ribbed plates in the green state.

Figure 3 $Y_2O_3 \cdot ZrO_2$ ceramic electrolyte cross flow SOFC stack.

Ceramic membrane-based thin chemical vapour deposition/ electrochemical vapour deposition (CVD/EVD) grown ceramic electrolyte layers for solid oxide fuel cells

K.J. de Vries, Y.S. Lin, L.G.J. de Haart, A.J. Burggraaf

Laboratory of Inorganic Chemistry, Materials Science and Catalysis
University of Twente, P.O.Box 217,7500 AE, Enschede, The Netherlands

ABSTRACT
Investigations have been performed to produce very thin (a few μm or less) gas-tight layers of yttria stabilized zirconia for SOFC applications in and on porous ceramic membrane layers by means of a modified CVD and EVD technique respectively. Simulation results of a mathematical modelling will be compared with experimental results.

1. INTRODUCTION

Solid oxide fuel cell (SOFC) reactors are attractive energy conversion systems. From model studies it is expected that they will have a high current efficiency, of about 60% (Steele 1987).
The applicability of the presently developed SOFC reactors can be improved when their main limitations can be avoided or at least decreased. The three largest sources of loss are: (i) The application of cubic yttria stabilized zirconia (YSZ) as electrolyte phase. The relatively low specific conductivity of this material requires, although relatively small (60-150 μm) electrolyte thicknesses are used, operating temperatures of at least 1000°C for having acceptable low bulk polarization losses; (ii) The high operating temperature of the SOFC reactor has the important consequence of a smaller OCV value of the solid oxide fuel cells compared to the OCV value at lower temperatures; (iii) The usually applied oxygen electrode materials are perovskites, e.g. Sr substituted $LaMnO_3$. These kinds of perovskites have a relatively low oxygen vacancy concentration at the P_{O_2} conditions at the oxygen electrode. Therefore these materials require to have a high open porosity when applied as oxygen electrode material in order to let pass oxygen from the atmosphere to the three-phase boundary area formed by the intercepts of the electrode material pores and the electolyte phase. The consequences are (i) a very severe constraint in the interface area at the electrode/electrolyte which is available for transfer of oxygen through this part of the cells. Moreover the required transport of oxygen through the pores of the oxygen electrode is a limitation too.
In this paper we present a technique together with a comparison between modelling results and some experimental results, for the production of (very) thin supported electrolyte layers of YSZ.
For SOFC applications the (very) thin YSZ layers that we produce by a modified CVD and EVD technique respectively, should have been deposited in and on a porous (perovskite)oxygen electrode layer with membrane top-layer properties. In this paper however stress was laid on (very) thin

YSZ layer production aspects. In this study we have applied α-Al$_2$O$_3$ as open porous membrane layer system in and on to which YSZ was grown.

The production of YSZ deposits for SOFC by a modified CVD and a EVD technique respectively was published before (Isenberg 1977; Dietrich et al., 1984; Carolan et al.,1987). A thorough mathematical modelling study of the modified CVD phase of YSZ, taking into account gas phase diffusion, convection and chemical reaction between the reactants (vaporized ZrCl$_4$,YCl$_3$ and H$_2$O)in the (membrane) pores was published by Lin et al. (1989).

2.   THEORETICAL AND EXPERIMENTAL

Figure 1 shows a cross-section of the schematic of an open porous ceramic support with large pore-diameters, covered by an open porous ceramic membrane layer. Both support and membrane layer are in this paper of α-AL$_2$O$_3$. Characteristic for a membrane layer is that its porosity is large (50%), the layer thickness is (very) small and the mean pore size is (considerable) below 1μm, whereas the pore size distribution is also rather small.

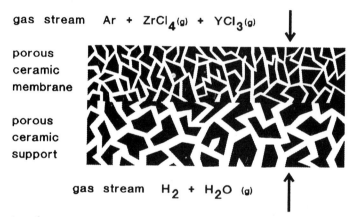

gas stream   Ar + ZrCl$_4$(g) + YCl$_3$(g)

porous ceramic membrane

porous ceramic support

gas stream   H$_2$ + H$_2$O (g)

Fig. 1.Cross-section of the schematic of an open porous
        ceramic (α-Al$_2$O$_3$) support wth large pores
        covered by an open porous ceramic (α-Al$_2$O$_3$)mem-
        brane layer.

The mean pore size in the presently described experiments is 0.16μm. Characteristic for the modified CVD and the EVD technique is the separated supply of the reactants (i) (vaporized) metalchloride mixture of ZrCl$_4$ and YCl$_3$ and (ii) watervapour-hydrogen mixtures, each from one side into and to respectively the porous membrane. In the CVD phase the reactants react in the membrane pores according to equation (1). After pore closure deposition proceeds by Electrochemical Vapour Deposition (EVD), unless the deposited material has at least some mixed (oxygen ion ànd electron) conducting properties, YSZ is just enough mixed conducting, according to equation (2).

$$(1-x)ZrCl_4 \; + \; xYCl_3 \; + \; (2-x/2)H_2O \; \longrightarrow \; Zr_{(1-x)}Y_xO_{2-x/2}\downarrow \; + \; (4-x)HCl\nearrow \tag{1}$$

At the water side of the closed pore (See Figure 2): Equation (2a)
$$2H_2O \; + \; 4e^- \; \longrightarrow \; 2O^{2-} \; + \; 2H_2\nearrow \tag{2a}$$

At the metalchloride mixture side of the closed pore the YSZ deposition
process continues according to equation (2b), which is made possible by
the electrochemical diffusion of oxygen ions from the watervapour side to
the metalchloride side of the closed pore, under the driving force of the
oxygen activity difference.

$$(1-x)ZrCl_4 + xYCl_3 + (4-x)/2H_2 + (2-x/2)/2O_2 \longrightarrow$$

$$Zr_{(1-x)}Y_xO_{(2-x/2)} \downarrow + (4-x)HCl \nearrow \qquad (2b)$$

If a mixture of $ZrCl_4$ and of YCl_3 in a proper composition is taken as one
reactant a solid oxide conducting thin layer of e.g. YSZ17 is produced.
The solid layer is deposited on the membrane pore walls. If the CVD
process is properly controlled, the modified porous membrane will get one
of the following three forms, mentioned in Figure 2 as 1, 2 and 3 respectively. In 1 the inner pore walls of the original α-Al_2O_3 ceramic membrane
is covered by a very thin deposited layer of YSZ, thus modifying the
physical and chemical properties of the inner membrane pore wall. In 2,
the pore-diameter is narrowed to any desired size. Important for SOFC is
that the pore-narrowing is conducted to take place near the entrance side
of the metalchloride mixture into the membrane pores. In 3, the pore is
just closed. The system is now gas-tight and is the equivalent of a very
thin semi-permeable or permselective "wall". CVD phase is blocked now.
Layer growth can continue now only at the metalchloride side of the closed
pore by the EVD phase, as described before. The latter gives a (very) thin
(YSZ) layer and is presented as point 4 in Figure 2.

gas stream   Ar + MeCl$_x$

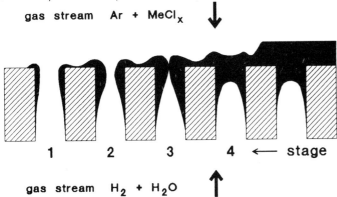

gas stream   H$_2$ + H$_2$O

Fig. 2. Exploded view of a part at the surface of the
schematic of the membrane toplayer. 1,2 and 3 correspond
to different products of modified CVD of the membrane.
4 corresponds to an EVD grown (very) thin (YSZ) layer.

2.1   Kinetics

The kinetics of the CVD phase involves the following steps: (i) interdif-
fusion of the two reactants into the membrane pores; (ii) chemical reaction and deposition of the solid product on the membrane pore wall; (iii)
diffusion of th gas product (HCl) out of the pores of the composite layer
system. The CVD process can be described (Lin et al. 1989) by a mathematical model, which takes into account diffusion, convection and chemical
reaction inside the pores. The resulting mathematical model consists of a

set of ordinary differential equations and the solution gives the conditions of the YSZ deposit distribution across the membrane pore. The simulation results (Lin et al. 1989) show that the deposit distribution is determined by reaction kinetics, the diffusivity of the metalchloride mixture and by several experimental conditions such as membrane pore dimensions, bulk phase concentrations of the two reactants, membrane temperature and the total pressure drop over the supported membrane layer. The two first mentioned parameters are combined in the Thiele module Φ, which is proportional to the ratio of the reaction rate constant (K) and to the effective pore diffusivity of the metalchloride mixture (D_m^O):

$$\Phi = \frac{K \, L^2}{D_m^O \, R_O} \, (C_m^O)^{M+N-1} \qquad (3)$$

Here L and R_O are the membrane layer thickness and the initial membrane pore radius, respectively. C_m^O is the metalchloride bulk concentration and M and N are the reaction orders on water and metalchloride, respectively. Figure 3 shows the simulated deposition profiles for different values of the Thiele module for the case of reaction order one on metalchlorides and zero on water. It is obvious that increasing the Thiele module gives deposition profiles which are more pronounced towards the side from where the metalchloride mixture is carried into the membrane pores and also deposition rate is faster.

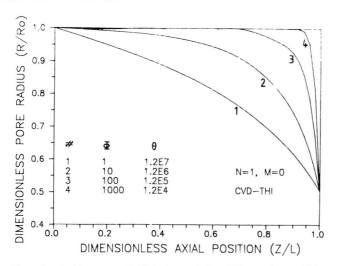

Fig. 3. Influence of Thiele module (Φ) on deposition profiles for the reaction order one on metalchloride and zero on water.

It is obvious that for the latter combination of reaction orders the higher the Thiele module the more the maximum deposition is located near the edge of the membrane layer exposed to the metal chloride mixture. From these results of the mathematical modelling it can be concluded that the experimental conditions must be selected in such a way as to obtain a reaction order of one on metalchloride and zero on water and a high Thiele module. In that case the CVD process yields a pore closure which is located at the side where the metalchloride vapours enter the porous membrane and the YSZ deposit penetrates only slightly into the membrane.

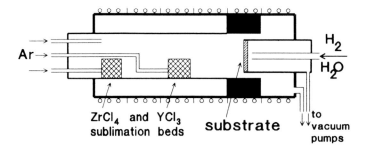

Fig. 4. Experimental setup of the hot-wall type of reactor used in the CVD and EVD phases. The substrate is provided with coarse pores. The substrate will in practice be covered by a porous membrane toplayer for carrying out the depositions.

Figure 4 depicts the reactor system in which CVD and EVD based layer productions takes place. For generation of well defined and controlled partial vapour concentrations of the metalchlorides seperated sublimation systems are required. The conditions used during CVD and EVD processing are 1000°C and a total pressure of a few mbar. Argon was applied as carrier gas also for carrying watervapour-hydrogen mixtures.

3.   RESULTS AND DISCUSSION

In the membrane toplayer of $\alpha-Al_2O_3$ that is applied in these investigations the ratio of the pore diameter to the pore length is rather small. This is in favour of a rather large Thiele module as discussed by Lin et al. (1989). This means that from theory, deposition of the metalchloride phase, e.g. YSZ, should take place near the membrane surface which is exposed to the metalchloride vapour mixture.
Figure 5 shows an EDS profile-analysis of a membrane sample that was cut through parallel with the direction of the diffusion of both kinds of reactants in the membrane pores. The horizontal scale in Figure 5 corresponds to a thickness of the membrane layer of 20µm. The right hand side of the scale being the edge of the membrane that was exposed to the metalchloride vapour. Indeed the EDS data indicate that the deposit of the metaloxides, e.g. YSZ, is only present in the range of a few micrometers thick in the membrane pores from the metalchloride side. Obviously the amount of deposited YSZ in the membrane pores is found to increase towards towards the metal chloride side of the membrane.

The ratio of yttria/zirconia in the deposit corresponds to about 8.5mol%. This ratio is apparently determined by the ratio of yttriumchloride/ zirconiumchloride in the gas phase.
By in-situ measurements of the gas-permeability at intervals of five minutes, the pore diameter decrease was followed to the moment of pore closure. For the presently applied porous membrane and experimental conditions this took about 20 minutes. Continuation of the deposition of YSZ conducted by the EVD phase for about two hours, without changing experimental conditions, resulted in an YSZ layer on top of the initially

porous membrane of about 8µm, as shown in Figure 6.

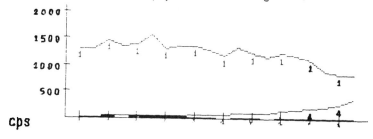

measured at 20 points

Fig. 5. EDS profile analysis of cross-section of a membrane layer subjected to the CVD phase of YSZ in a direction parallel to the diffusion of both kinds of reactants. 1 = Al, 4 = Zr. Total scale length 20µm.

Fig. 6. SEM photograph of a cross-section of a CVD/EVD deposit of YSZ in and on a porous alumina membrane.

4. CONCLUSIONS

1. A unique CVD and EVD technique is presented for the production of (very) thin electrolyte layers of YSZ for SOFC reactors, by modifying porous membrane layers on top of a porous substrate (oxygen electrode

2. The mathematical modelling results provide a Thiele module rule that can predict location of pore closure location inside the membrane pores. Experimental CVD/EVD results are shown which agree well with the theoretically developed parameter dependence.

5. ACKNOWLEDGEMENT

This investigation was supported by the Dutch Ministry of Economical Affairs under grant: IOP Technical Ceramics No. 87 A045.

6. REFERENCES

Carolan M.F., Michaels J.M. 1987 Solid State Ionics 25 pp 207-16
Dietrich G., Schäfer W. 1984 Int. J. Hydrogen Energy 9 pp747-53
Isenburg A.O. 1977 ECS Symp. Electrode Materials, Processes for Energy
    Conversion and Storage Vol 77-6 pp 572-81
Lin Y.S., De Vries K.J., Burggraaf A.J. 1989 Submitted to J. de Phys., in
    press
Steele B.C.H. 1987 "Ceramic Electrochemical Reactors, Current Status and
    Applications" Ceramionics, Surrey (UK) pp 21-32

The mechanical performance of ceramic dust filter elements in the tertiary dust capture filter of the Grimethorpe pressurised fluidised bed combustor (PFBC)

R Morrell[1], D M Butterfield[1], D J Clinton[1], P G Barratt[2], J E Oakey[3], G P Reed[3], M Durst[4], and G K Burnard[5]

[1] National Physical Laboratory, Teddington, Middlesex, TW11 0LW
[2] British Ceramic Research Ltd, Stoke-on-Trent, Staffs, ST4 7LQ
[3] Coal Research Establishment, Stoke Orchard, Cheltenham, Glos, GL52 4RZ
[4] Schumacher GmbH & Co KG, Zur Flugelau 70, D7180 Crailsheim, F R Germany
[5] formerly at Grimethorpe PFBC Establishment, Grimethorpe, S Yorks, S72 7AB

ABSTRACT: Silicon carbide filter elements used in the Grimethorpe PFBC hot gas filter have been examined mechanically and microstructurally after various periods of exposure to PFBC dust. Ultrasonic velocity of sound measurements correlated well with destructive mechanical test results. Many, but not all, of the elements suffered reductions in strength, and some showed some cracking along their bores. Microstructural examination revealed little reaction between the aluminosilicate bond phase and the PFBC dust, but demonstrated microcracking leading to loss of structural integrity. The pulse-cleaning system using cool pressurised air is concluded to have subjected the elements to excessive levels of thermal shock and fatigue.

1. INTRODUCTION

In the method of pressurised fluidised bed combustion (PFBC) for coal-fired power generation, both sulphur and NO_x emissions can be controlled, and at the same time the overall efficiency can be increased compared with conventional pulverised fuel combustion. The increased efficiency is achieved by use of combined-cycle operation incorporating a gas turbine in parallel with a steam generation circuit. In order to protect the turbine from erosion and corrosion by dust in the combustion exhaust, there is a need for a high-temperature, high-pressure (HTHP) gas clean-up system between the combustor and the gas turbine. Gas clean-up using cyclones, while sufficient for an acceptable turbine life, cannot meet proposed standards for dust emission to the atmosphere without further cleaning downstream of the gas turbine. There is therefore a strong incentive to develop higher efficiency clean-up methods capable of operating under the conditions seen between combustor and turbine, which are typically 10 atmospheres pressure at 800 °C. Barrier filtration is seen as having considerable potential for development. The US Electric Power Research Institute (EPRI) recognised the need for research and development work on such devices, and have collaborated with the Grimethorpe PFBC Establishment operated by British Coal and Central Electricity Generating Board in the UK, where the facility offered unique possibilities for such work.

A pilot-scale filter vessel, an order of magnitude greater in size than any filter previously tested under PFBC operating conditions, was constructed for test on part of the off-gas (up to 7 kg s^{-1}) from the PFBC. This vessel was of sufficient size that statistically significant conclusions could be drawn about the design, operation, and reliability of the filter (Reed et al.(1987), Tassicker et al.(1989)). A schematic diagram of the filter vessel is shown in Fig.1. A large area of filter surface was provided by more than 100 1.5 m long porous silicon

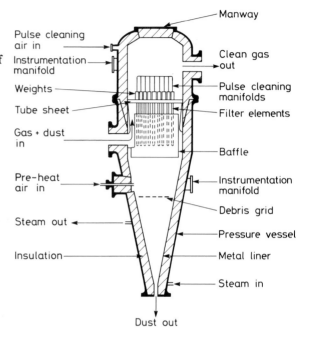

Fig.1 Schematic of design of Grimethorpe filter vessel.

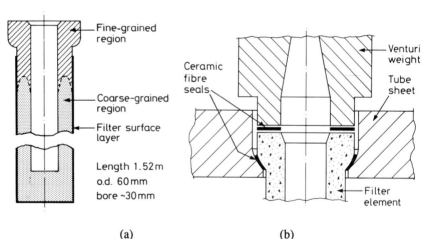

(a)   (b)

Fig.2 Silicon carbide filter element (Dia-Schumalith F40) showing (a) its make-up and dimensions, and (b) the method of mounting in the filter vessel.

carbide tubular filter elements (Dia-Schumalith F40 from Schumacher GmbH) closed at one end and with a flange at the other (Fig 2a). These were suspended from counterbored holes in a horizontal steel plate (the tubesheet) inside the filter vessel as shown in Fig.2b, each weighted with a 10 kg tubular steel mass to prevent the pressure differential of up to 300 mb from lifting the filters. Ceramic-fibre gaskets were placed between each element and the tubesheet, and between the weights and the elements to provide a gas seal and to avoid stress concentrations. The body of the filter material comprised a coarse-grained clay-bonded silicon carbide ceramic with wide, open-porous channels to permit gas flow, Fig.3. The outer surface was coated with a fine wash-coat of similar, but much finer-grained, lower-porosity material incorporating ceramic fibres to act as the filter barrier. The top of the filter and the flange comprised medium grain-size material of the same type to achieve higher strength in this region. The filtered dust deposit built up on the external surface of the filter elements, and periodically had to be removed. This was achieved by a reverse flow air-pulse cleaning system, operated intermittently, each element being cleaned about every two minutes. Air at a reservoir pressure of up to 45 atmospheres was directed down the open end of each element through a centrally placed nozzle by briefly opening and closing a solenoid valve connected between an array of such nozzles and the pressurised air reservoir. The bore of the retaining weight was shaped as a venturi to enhance the cleaning pulse. The dust cake accumulating on the filter surfaces would then be blown off, and would fall to the bottom of the filter vessel from where it was removed by an Archimedean screw arrangement. It should be noted that not all the elements were actively filtering and being cleaned during the campaigns.

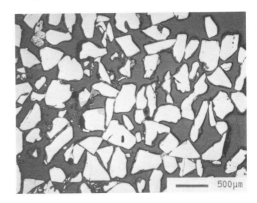

Fig.3 Microstructure of the coarse-grained body of a filter element.

500μm

A series of trials were performed using the filter with some differences in operating conditions, particularly with regard to the fuel used in the PFBC, to the lay-out of the filter elements in the tube-sheet, and to the pulse cleaning system. The purpose of this paper is to report the findings of the post-mortem analysis of filter elements removed after various periods of operation and to examine the extent to which their mechanical and microstructural integrity had changed as a consequence of the arduous operating conditions.

2. MATERIAL CHARACTERISATION

2.1 Ultrasonic examination

In order to assess the basic integrity of the filter elements, time-of-flight measurements were made using ultrasonic pulses transmitted in longitudinal mode along the full length of the elements. "Pundit" instruments (CNS Instruments) were used with 200 kHz (British Ceramic Research Ltd (BCRL) or 54 kHz (National Physical Laboratory (NPL)) transducers. The time- of- flight is related to the Young's modulus, and hence to the integrity of the bonding between grains. Rubber caps were used on the transducers to ensure good coupling to the filter element. The technique could also be used on short broken or cut lengths of element.

2.2 Mechanical testing

Of the various test methods that could have been employed to evaluate strength properties, it was decided that the simplest ones using ring specimens were most appropriate. Three techniques were used: full-ring diametral compression (DC) (Bortz and Lund (1961)), C-ring diametral compression (CC) and C-ring diametral tension (CT) (Davidge (1981). Rings were cut dry using a steel-bonded diamond saw, and C-shapes were made by slotting the ring radially by hand. The compression specimens were tested between sheets of card in an Instron mechanical testing machine equipped with compression platens. The tension specimens were tested in a purpose-built lever jig, again using the Instron machine in compression.

The DC and CT tests place the maximum tensile stress on the inner surface of the specimen, while in the CC test, the outer surface of the specimen is subject to the maximum tensile stress. The C-ring specimens can thus be used to test separately the two surfaces of the filter element. Testing of beams cut parallel to the length of element was discounted as a routine technique because an excessive amount of machining was required, coupled with difficulties in retaining knowledge of their orientation relative to the original element surface. The equations used to calculate fracture stress σ_f were:

Bortz ring:
$$\sigma_f = \frac{Fk}{\pi BR_2}$$

C-ring tension:
$$\sigma_f = \frac{((R_2^2 - R_1^2)/2T)((T/\log R_2/R_1) - R_1)F}{((R_2^2 - R_1^2)/2T) - T/\log R_2/R_1)BTR_1} - \frac{F}{BT}$$

C-ring compression:
$$\sigma_f = \frac{((R_2^2 - R_1^2)/2T)(R_2 - T/\log R_2/R_1)F}{((R_2^2 - R_1^2)/2T) - T/\log R_2/R_1)BTR_2} + \frac{F}{BT}$$

where F = fracture load, B = ring thickness, R_1, R_2 = inner, outer ring diameter respectively, $T = R_2 - R_1$, and k is a constant depending on the ratio R_2/R_1. The value of k taken was 20.3 for $R_2 = 2R_1$ using the original analysis (Ripperger and Davies (1947)). Small variations in R_2/R_1 were found between specimens, up to 5%, but since the specimen bores were neither central nor circular, and the probable errors were not significant compared with the scatter of individual results, values of k were not recalculated for each ring.

A variety of other standard techniques were used including X-radiography, density and porosity tests, Young's modulus measurements, thermal shock tests, thermal expansion tests, hot strength tests and simple creep tests.

2.3 Visual and microstructural examination

Visual examination of elements and specimens cut from them proved to be invaluable in detecting features that could not otherwise be seen owing to the microstructural inhomogeneity. For example, cracks in the coarse grit structure could be revealed by their "grooving" when intersected by a diamond saw cut. Microstructural analysis was the most difficult aspect of the characterisation programme. It proved almost impossible to obtain damage-free polished sections of the filter material due to the low fracture toughness of SiC grit particles, despite attempts to support the edges of grit particles by the use of mounting resin impregnation. Problems were also encountered with differential polishing between grit particles and glassy bond phase. X-ray diffraction was also used to determine changes to the material, but was not found to be a particularly sensitive technique because of the small proportion of bond phase, making detection of changes in it difficult to resolve.

Scanning electron microscopy (SEM) was used to examine fracture surfaces of strength specimens, both before and after exposure in the filter, and energy-dispersive X-ray analysis (EDX) was used to examine interactions between PFBC dust deposits and the filter. Considerable care had to be taken with interpretation in case features observed had been introduced by specimen preparation techniques, especially fracturing specimens. A technique for observing fracture paths was devised whereby a specimen would be internally coated with silver using $AgNO_3$ and NH_4OH solutions. Subsequent observation of secondary electron SEM images would then allow the freshly fractured areas to be distinguished from the pre-existing surface.

3. TEST RESULTS

The complex series of experimental runs with the filter can be considered for the purposes of this paper as being equivalent to an experiment involving various periods of exposure of the elements to broadly similar dust filtering conditions at 10 atmospheres pressure and about 800 °C. The vessel was opened and the filter elements were removed after a short initial trial and after each of three test series of about 300, 300 and 190h respectively. Ultrasonic characterisation was used by BCRL on most elements after the test series, but unfortunately no data exist for them as initially installed. Selected elements from each test series were then subjected to destructive examination, with cumulative exposure periods of up to 790h.

3.1 Ultrasonic examination

Unused filter elements examined prior to their installation in substitution for those removed at earlier stages showed a narrow spread in end-to-end times-of-flight of 285-320 µs with the majority in the range 290-300 µs, indicating a good degree of consistency in manufactured density and degree of bonding. Measurements made over equally spaced shorter distances along the external cylindrical surface of the elements showed very little variation from one end of the element to the other.

After exposure, many elements showed a significant increase in times-of-flight, the range changing to 288-378 µs. Elements showing an increase in the first exposure period generally showed a smaller one in subsequent exposure, although some rather curiously showed decreased values in the third period. Attempts were made to try to correlate the figures with other factors, such as original batch number, position in the filter, and position in the pulse-cleaning manifold system, but no sensible conclusions could be reached. It was clear that some elements were being significantly changed by exposure whereas others were not.

In laboratory experiments, it was found that the dust loading had no effect on the time-of-flight of used and unused elements. Furthermore, the time-of-flight was unaltered by heat-soaking used elements at 800 °C for 200 h after service in the filter. Ultrasonic results should thus not be biased by dust loadings inside elements, and that changes detected in service are not reversible by subsequent heat-soak "healing" at the same temperature.

3.2 Strength tests

Specimens cut from the coarse-grained as-manufactured material near the element flange and near the closed end showed very similar strengths in all three ring tests, indicating little difference between the inside and outside surfaces, and also consistency along the element length. Mean values were typically 12 MN m^{-2} or higher. Fracture surfaces were fairly flat and there was very little loose debris produced during fracture. Specimens cut from exposed elements appeared to retain this level of strength when tested as complete rings, but both C-ring tests gave significantly lower results in many elements, with the CC tests (i.e. tensile stress on the external filter surface) giving higher results than the CT tests (i.e. tensile stress on the filter bore). In some cases, C-ring specimens were too weak to be tested; of those that could be tested, strength figures as low as 3 MN m^{-2} were obtained. Visual observations suggested that some specimens were already cracked before testing (Fig.4), always from the bore surface. The outer filter skin never showed any signs of cracks. Fracture surfaces were macroscopically rougher (Fig.5), and fracture did not always originate along the line of maximum stress in the specimen, but at random positions.

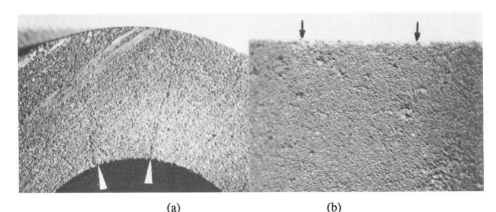

(a)   (b)

Fig.4 Cracking from the bore surface (fine-grained flange region) after service revealed by grooving during diamond saw cutting of a test specimen (a), but invisible on bore surface at corresponding positions (b).

(a)   (b)

Fig.5 Typical fracture surfaces of C-ring tension specimens of (a) an unused element, and (b) a used element.

One unused and two used elements were cut into CC specimens and into flexural strength test bars. Strengths recorded at 800 °C were similar to those recorded at room temperature, and there was little difference in result between the two types of test.

3.3 Correlation of strength and ultrasonic time-of-flight

In ceramic materials the ratio of strength to Young's modulus is usually about 0.001, and thus it was reasoned that in the present study there should be a good correlation between strength and ultrasonic time-of-flight. Because the BCRL ultrasonic data were on full-length filter elements incorporating the finer-grained denser flange region, fresh measurements were made by NPL on remaining uncut short sections of elements. Figs. 6, 7 and 8 show plots of time-of-flight per metre length of main filter body against strength for the three ring tests on all elements tested. While there is a correlation in all three cases, the most significant results were obtained with CC tests. The general behaviour clearly indicates that elements can

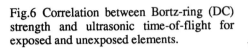

Fig.6 Correlation between Bortz-ring (DC) strength and ultrasonic time-of-flight for exposed and unexposed elements.

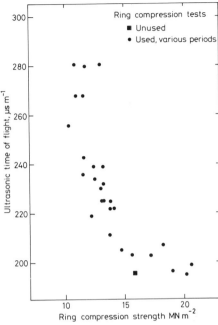

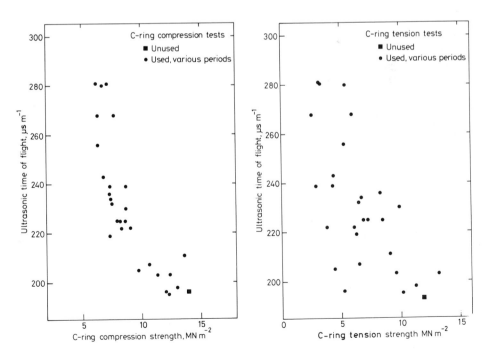

Fig.7 As Fig.6 but C-ring compression (CC).   Fig.8 As Fig.6 but C-ring tension (CT).

weaken and loose stiffness, with greater weakening at the bore surface (CT tests) than at the outside surface (CC tests). This corresponds to the significant cracking observed from the bore surface, but in addition, there also appears to be a general weakening of the outer zones of material.

Fig.9 Fracture surface of element with 300 h exposure with silver coating applied to the bond surface before fracturing. Fractured areas appear dark.

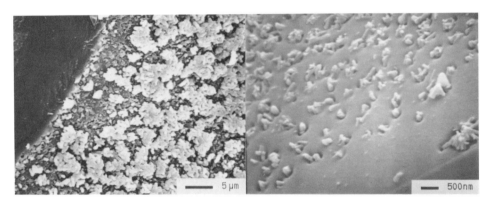

Fig.10 Adherent dust on the bond surface after 790 h service, exposed by fracturing a specimen.

Fig.11 Minor interaction between dust particles and the bond surface after 300 h exposure.

3.4 Microstructural observations

X-ray and SEM observations showed that the SiC grit particle surfaces were completely covered in an aluminosilicate glassy bond that produced necks between particles. The bond had a small amount of crystallinity in the as-manufactured condition, and this tended to increase slightly with exposure to the filter vessel operating temperature.

Examination of fracture surfaces of unused elements showed that a substantial proportion of obvious fracture area did not comprise solely glassy bond necks between grit particles, but was in fact SiC grit. This was confirmed by using EDX analysis on silver-coated and fractured specimens. Large areas could be seen in which aluminium (which is present in the bond phase) and silver were absent (Fig. 9). After service in the filter, a greater proportion of SiC was evident, especially in the weaker elements.

All bond surfaces were decorated with dust which had passed through the external filtering layer, but none of the filters was clogged in any way. The dust comprised individual particles of anhydrite ($CaSO_4$) as major species, with smaller amounts of haematite (Fe_2O_3) and α-quartz (SiO_2), and these could be identified as discrete particles 0.1 to 1 μm across on the bond surface (Fig. 10). The particles appeared to have adhered to the bond surface, and reaction had occurred as clearly seen by the formation of small craters around particles, indicating possible reaction zones (Fig.11). In some cases, "drilling" to a depth of about 1 μm had occurred, but in polished or fractured cross-sections, no evidence could be found to suggest that this had caused major changes to the bulk of the bond which remained substantially glassy up to 790 h operating time.

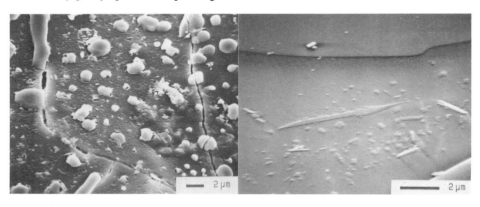

(a)   (b)

Fig.12 Microcracks in (a) bond external surface and (b) at the bond/grit interface in exposed elements, the latter revealed by light HF etching a polished surface with grit particle at top.

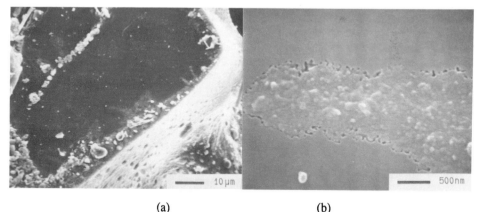

(a)   (b)

Fig.13 SEM images of a silicon carbide grit fracture surface prepared after exposure showing (a) dust penetration, and (b) possible oxidation of the surface.

Close examination of fractures in the SEM revealed a number of differences between unused and exposed elements. In used filters, microcracks could occasionally be seen running along or near the interface between grit and bond (Fig.12). The grit fracture surfaces were not clean cleavage planes in all cases (Fig.13). Some showed fine-textured features which appeared to contain no EDX-detectable species apart from Si (i.e. that they were either SiC or SiO_2), suggesting that slight oxidation may have occurred, presumably because a crack through the grit was developing whilst in service.

3.5 Other test results

A few X-radiography tests were made using film in cylindrical form along the bore of the elements and through whole elements. The inhomogeneity of the microstructure masked any damage that might have accumulated in service. Some banding due to density variations could be seen in both unused and used elements. Only on one occasion could a crack be detected, but this was visible to the unaided eye.

Density measured by immersion was found to vary along the length of an element by typically $\pm 1\%$, limiting the possibility using such measurements to resolve changes due to exposure, assuming that dust loadings were negligible.

Thermal expansion measurements were made on unused and used element material to determine whether changes to the structure had occurred, particularly crystallisation of silica phases, the phase changes of which might lead to the development of microcracks. No differences were detected. A sample of the original unfired bond material was pressed into a bar and fired to $1200\,°C$. Its thermal expansion characteristics gave a slightly higher figure than that for the SiC elements, and showed a small inflection at $573\,°C$, corresponding to the presence of a tiny amount of residual α-quartz. At lower temperatures, there was no evidence of a cristobalite transition which might have caused mechanical weakening on thermal cycling. High-temperature measurements showed signs of softening of the bond only at temperatures of $900\,°C$ and above, well above the filter operating temperature.

Simple flexural creep tests at a maximum nominal stress of about $1\,MN\,m^{-2}$ were made on used and unused specimens. In the unused material no significant curvature could be produced up to $1100\,°C$ with exposure periods of $100\,h$ or more. In the used elements, curvature was detected at $900\,°C$ after $100\,h$, and the specimens became extremely fragile after the test, suggesting that the deformation was essentially due to the growth of pre-existing cracks.

Because pulse cleaning of the elements must involve some thermal shock, the behaviour of the material under shock conditions was considered important. Tests on bar specimens by the manufacturer showed no significant change in Young's modulus or flexural strength with repeated air-jet quenching from $900\,°C$ to room temperature. Under more-severe water-quench conditions, both Young's modulus and strength were reduced to about one-third of their original values after two cycles, but then was unchanged up to 16 cycles. Simulation of quenching the bore of an element using a cold air blast along a section of element heated at $900\,°C$ produced no change in subsequently measured CT tests.

4. DISCUSSION

The results of the post-mortem tests on the filter elements revealed that for the most part they had fulfilled their function in removing fine particles from the PFBC. However, it is clear that the strength of the element material was being reduced in a somewhat unpredictable manner by exposure. In some elements it was barely affected; in others it was reduced in a major way. Visual examination showed that C-ring specimens of weak elements tended to be cracked from the bore. However, the ultrasonic time-of-flight measurements are expected to be substantially unaffected by such cracks as the pulses would propagate by the fastest route, which would be through uncracked regions near the outer surface. The facts that the best strength correlation with time-of-flight measurements was in the C-ring compression tests, and that these strength results were all consistently much greater than the C-ring tension results but were still less than those of unused elements, suggest that there is a general weakening of the microstructure as well as the obvious bore cracking.

The cause of the bore crack development is uncertain but is likely to be due to the thermal shock of the cleaning pulse of relatively cold air (at about 250 °C). Measured temperatures in element bores during pulse cleaning were found to drop rapidly by up to 200 K. The used elements had experienced up to 10,000 such pulses. The intensity of the pulse was much greater than that imposed by the laboratory air-quenching tests. However, since the array of pulse cleaning nozzles had different lengths of supply pipe and were at different and relatively uncontrolled positions in the venturi weights as a result of thermal movements after installation, it is highly likely that different elements experienced different degrees of thermal shock. The decrease of strength and stiffness (from the time-of-flight results) of the outer regions of the element wall may also be due to the same cause, but a different mechanism. The passage of cold air through the porous structure is likely to cause localised thermal stresses between the bond phase and the underlying SiC grit. This would tend to cause the development of any pre-existing cracks or defects, especially along weakly bonded interfaces or cleavage planes. Such microcracks were observed in the SEM examination. However, it remains unclear whether the large cracks at the bore develop as a single event directly as a result of rapid quenching of the bore relative to the outside of the element, or originate from linking of microcracked areas in a progressive way as the structure fatigues. The interaction between the dust contamination and the bond phase in the high-pressure water-bearing atmosphere is also not entirely clear. Further work needs to be done using longer exposure periods to ascertain the significance of the slight interactions seen. Certainly elements which were exposed to the PFBC environment in the vessel but which were not being actively cleaned by air pulsing suffered far less strength degradation than those actively filtering and being cleaned.

From this study, a number of conclusions can be drawn about the suitability of the filter elements for long-term service in a full-scale filter vessel for commercial use. Firstly, it is clear that 790 h total exposure to the PFBC, although not an adequate duration to prove the system for economic production campaigns, was successfully survived by the Dia-schumalith elements with only minor interaction with the PFBC dust. Secondly, although the elements were hanging vertically and were subjected to stresses due to their own mass, the mass of accumulating dust deposits, and the pressures and thermal shock of pulse cleaning, they survived the arduous service conditions. Thirdly, because of the measured loss of strength in service, consideration has to be given to reducing the thermal shock effects of pulse cleaning to a minimum consistent with cleaning the elements. If feasible, the air supply used could be heated to near the vessel operating temperature. Fourthly, further to establishing the likely degradation mechanisms, controlled pulse-cleaning experiments need to be undertaken. Lastly, changes to the basic filter material may be desirable if other properties are not adversely affected. For example, the expansion coefficient mismatch between bond and SiC could be reduced to reduce local stresses, or the bond content could be enhanced to increase strength, or the grit size of the silicon carbide could be reduced at the expense of an increase in pressure drop across the filter.

The value of ultrasonic characterisation has been fully demonstrated as a non-destructive means of monitoring element characteristics. If possible, in-situ monitoring of elements in a filter would be highly desirable in order to track changes in mechanical performance without removing them from the filter. Clear correlations exist between the ultrasonic times-of-flight and the strengths of the elements. Of the different strength tests employed, the C-ring tests have shown themselves to be the simplest yet most valuable and economic, and are capable of following strength changes of both internal and external surfaces of the element. The full ring test is less successful because although it should test the bore surface, if this is already cracked, the specimen effectively acts as a double C-ring compression specimen, and the peak load corresponds to twice the external surface strength, a somewhat misleading situation. Beam tests are much less convenient from the specimen preparation point of view, and it is easy to lose orientation relative to the original element geometry.

Although the microstructural analysis provided valuable insight to the possible degradation mechanisms, its difficulty must not be under-emphasised. The highly inhomogeneous, porous microstructure makes observation difficult, and conclusions about changes to it are difficult to draw. The relative importance of changes in bond crystallinity and of attack by dust species needs to be carefully examined in tests of longer duration than those of this pilot trial.

5. CONCLUSIONS

A post-mortem analysis of Dia-schumalith F40 porous silicon carbide tubular filter elements from the Grimethorpe PFBC dust filter vessel has shown that:
(1) the clay-bonded silicon carbide resisted the service conditions adequately from a dust contamination point of view for the maximum exposure period of 790 h;
(2) a significant number of elements suffered losses in stiffness, as determined from ultrasonic time-of-flight measurements, and in strength, as determined by full-ring and C-ring mechanical tests, while others did not;
(3) it is likely that excessive thermal cycling occurred due to the periodic pulse cleaning process using cool air, resulting in major cracking along the bores of some elements, and in general microcracking in SiC grit particles and between them and the bond phase;
(4) valuable experience has been gained in the engineering use of ceramic filter elements and a number of conclusions can be drawn for future developments;
(5) non-destructive evaluation of filter elements is a highly desirable step at all stages of monitoring mechanical performance.

ACKNOWLEDGEMENTS

This work was supported by the Electric Power Research Institute, Palo Alto, California, USA; Project Manager: Dr J Stringer. This task was coordinated by the Coal Research Establishment of British Coal. The authors wish to thank EPRI and the Director of the Coal Research Establishment for their permission to publish this paper. Dr Stringer is thanked for the useful discussions held and for reviewing this paper. The views expressed in this paper are those of the authors, and not necessarily those of the parent organisations.

REFERENCES

Bortz, S A, Lund, H H (1961), in Mechanical Properties of Engineering Ceramics, ed. Kriegel, W W, Palmour, H, (New York, Interscience Publishers) pp 383-404.

Davidge, R W (1981), private communciation.

Reed, G P, Burnard, G K, et al.(1987), Proc. Int. Conf. Advanced Coal Plant Technology and Hot Gas Cleaning, Dusseldorf, F R Germany, 2-4 Dec. 1987.

Ripperger, E A, Davies, N (1947), Trans.Amer.Soc.Civil Engrs. 112, Paper no 2308, 619-27.

Tassicker, O J, Burnard, G K (1989), Proc.Conf. Fluidised Bed Combustion, San Francisco, Calif. USA, 30 April - 3 May 1989.

Ceramic heat exchangers for domestic and industrial application

J. Heinrich*, J. Huber*, H. Schelter*, R. Ganz**, O. Heinz**
* Hoechst CeramTec AG, Werk Selb, Wilhelmstr. 14, 8672
  Selb, FRG
** Hoechst AG, FTT Neue Technologien, Postfach 80 03 20,
   6230 Frankfurt/Main, FRG

ABSTRACT
Heat exchangers for use in domestic heating devices, in
industrial heat transfer systems as well as in chemcial
engineering applications are made of silicon infiltrated
silicon carbide (SiSiC) because of its high temperature
strength and its corrosion resistance within a large pH
range. The main item of the heat exchanger systems presented
in this paper is a SiSiC element manufactured by tape
casting thin ceramic layers, laminating punched layers
together and sintering them to a monolithic gastight part.
Besides the processing technique, which can lead to a lot of
different geometries, the connection of the elements to each
other and to the peripheric housings is discussed.
In the field of domestic heating devices compact and condensing heaters can be built. As no corrosion takes place no
poisonous heavy metals are set free by the acid condensate
and thus a high lifetime is guaranteed. In chemical
engineering the heat exchangers can be used to raise the
concentration of acids and to recover heat from highly concentrated hot acids. In the area of industrial heat recovery
the flue gases of industrial furnaces can be used to preheat
combustion air. Corrosive flue gases of thermal cleaning
systems for exhaust air can be preheated and exhaust gases
can be cooled down below the acid dew point. While energy is
produced heat can be transmitted to another medium. Examples
are heat transfer systems during the cleaning of fumes or
the production of vapour. The application of single heat
exchanger elements, heat exchanger systems and their
technical data will be discussed.

1. INTRODUCTION

There are two ways in which ceramic materials may be of
interest for the design of heat exchangers. In the
high-temperature range above 800°C new opportunities open up
for the recovery of heat energy from waste heat in
industrial processes. On the other hand the high corrosion
resistance of ceramic materials offers alternative possibilities in the low-temperature range as condensing heat

exchangers. Increasing the service life and carrying out processes which are not possible or not economic with conventional materials are further arguments in favour of the application of ceramics in heat exchanger design (1). Opportunities lie, for example, in widening the temperature limits of metals, graphite, of PTFE and glass.

Because of its 100% leaktightness and its corrosion resistance, silicon infiltrated silicon carbide (SiSiC) is of particular importance. In contrast to metals no significant reduction in strength occurs for SiSiC up to approx. 1400°C (fig. 1). On the other hand ceramic materials are by nature brittle. At the design stage it is necessary, for example, to prevent tensile stresses, avoid point applications for loads and where possible convert forces into compressive stresses. The ceramic production technique permits only relatively limited component sizes. In the production of functional units for industrial applications, complex joining and connecting techniques are therefore necessary. The ceramic material silicon infiltrated silicon carbide (SiSiC) permits in combination with the tape casting technique the design of compact heat exchangers with highly complex structures. In many cases, flue gas streams contain particles, e.g. in forging furnaces. A cleaning facility must be available for dust-loaden flue gas streams. With liquid/liquid and liquid/gas heat exchange it may be meaningful to alter transfer surface, install turbulance generators etc. and hereby establish process optimizing designs. The tape casting technique described here offers the possibility for doing so and provides the design with greater opportunities than with any other ceramics production technique.

This paper will describe the efficiency, application and production of ceramic compact heat exchangers made from SiSiC. This is first a plate-type cross-flow design with external dimensions of 300 x 300 x 150 mm and rectangular channels. The appropriate function for this type of element is the preheating of combustion air from corrosive flue gases with temperatures above 800°C. When metal heat exchangers are used, the hot exhaust gases are generally quenched by air to the permissible heat exchanger inlet temperature of approx. 800°C, thus destroying a high energy potential. The utilization of exhaust gas energy between 800 and 1350°C permits additional fuel savings of the order of 15-20% (fig. 2).
The second design described in this paper shall be used for a condensing heat exchanger. In order to obtain specific information on the efficiency of these ceramic compact heat exchangers, they must be tested under service conditions.

In a specially equipped test facility the heat transfer characteristics of the cross-flow heat exchanger elements have been determined as a function of various parameters. Results with gas/gas heat exchangers are shown and illustrative examples are given for the use of ceramic heat exchangers in different fields of industrial applications.

In another example the applications of a SiSiC counter flow heat exchanger in a domestic condensing boiler is described. Information data of a 20 kW unit are presented.

2. PROCESSING TECHNIQUE FOR CERAMIC HEAT EXCHANGER ELEMENTS

For the production of ceramic heat exchangers with a high specific heat transfer surface tape casting is a convenient molding process. This technique permits the production of very thin, flat, large-area tapes and based on this, complex structures. Essentially, the tape casting process consists of the suspending of fine powders in organic or aqueous solvents with the use of binders and plasticizers and the casting of these slips on a moving surface. After the vaporization of the solvents according to the binder systems a more or less flexible tape remains, which can be cut, punched or laminated (fig. 3). A summary of the huge range of possible systems for aqueous and nonaqueous solvents can be found in (2). A detailed description of rheological properties and the structure of organic materials for the process of tape casting is to be seen in (3). In contrast to a lot of other powder metallurgical processes sheets with a thickness between 0,2 and 1,5 mm can be produced by tape casting. Originally, this process was developed for the production of different electronical components, such as capacitor dielectrics, piezoelectrics, ferrites, substrates and multilayer packages (4). Because of the large surfaces obtainable by means of laminating or winding embossed or punched tapes, substrates for catalysts (5) and heat exchangers can be produced, too.

For heat exchangers produced from silicon infiltrated silicon carbide, the starting material is SiC-powder. After the molding and the burning out of the organic components the remaining pores are filled with silicon. This can either be effected through the gasphase or through the liquid phase. During this process no appreciable change in the overall dimensions can be observed. SiSiC is a dense material consisting of about 90% silicon carbide and 10% silicon. A detailed description of the different processing steps is given in (6).

3. DESIGN OF HEAT EXCHANGER ELEMENTS

Fig. 4 shows the construction of a typical heat exchanger element with rectangular channels in a cross-flow system. The heating unit is fabricated by laminating together two rippled basis plates. The media flow cross-wise through the block element.

For chemical appliances it is sometimes better to have a variable proportion of the areas of both sides of a heat exchanger. On the other hand round channels can be cleaned easier than rectangular channels. The advantage of a round tube is also a higher bursting pressure. Those points can be achieved, if the heat exchanger element is not fabricated by plates and stripes, but by laminating together punched cards. This technique enables the designer to produce various complex structures in one block element. Fig. 5 shows that by laminating together those different cards

vertically to the card area tube shaped channels are formed
for the one media. Additionally in this card gaps are fixed
in a way that by laminating together the cards in two levels
a further gap is formed so that the second media flows
around those "pipes" through the block element. This can be
described as a rippled pipe bundle in one block. (7).

In fig. 6 the two different types are shown. The external
dimensions of both blocks are the same, i.e. 300x300x150 mm.
The left element with the rectangular channels has a heat
exchange area of 1,5m^2 on both sides, while the right
element has an area of 0,52m^2 on the tube side and an area
of 2,27m^2 on the slit side. In case both constructions are
combined, elements with linear channels are achieved on both
sides. One product stream flows in a rectangular channel,
the other one in any other channel shape.

Fig. 7 shows as example a rippled basis plate for a
gas/liquid element. For fabricating the block element the
basis plates are laminated together in the same direction.
The liquid flows through the channels and reaches the gas in
the cross-stream. In this case an open cross-stream block
element with appr. 4m^2 gas- and 0,6m^2 liquid touched area
results. The connections for the liquid and the distribution
pipes can also be integrated into the ceramic block element.

Fig. 8 shows the same element type in 3 different sizes. The
connections for the liquid as well as the distribution pipes
are integrated into the element.

For domestic application the boiler part is a ceramic
gas-liquid counterflow heat exchanger (fig. 9). It is also
made of SiSiC and fabricated by an assembly of ceramic tapes
of 0.8 mm thickness and different stripes between the tapes.
The flue gas channels and the water channels are straight in
the major part of the heat exchanger. Near the water outlet,
e.g. the entrance of the hot combustion gas an intensive
heat transfer from the walls to the heating water is
achieved by the cross-flow pattern. Thus the maximum
material temperature will not exceed about 150°C. Some
typical heat exchanger data are shown in table 1.

The layout of this heat exchanger has been developped in
cooperation with the KFA Jülich (8).

4. CHARACTERISTICS AND PERFORMANCE DATA
Silicon infiltrated silicon carbide (SiSiC) is a composite
material. Silicon is integrated in the SiC microstructure,
the material has no porosity.

For the application in chemical apparatus SiSiC was tested
as to its suitability in some special media (table 2). The
conditions chosen normally lead to severe corrosion when
metals are used. The yearly weight loss in mm is calculated
out of two-week short-time tests. Here the excellent
corrosion resistance of SiSiC against aggressive media is
shown. Only hydrofluoric acid and a 50% NaOH affect SiSiC
essentially.

The material characteristics of tape cast SiSiC are described in table 3. The low density, the low coefficient of thermal expansion compared with steel and the high thermal conductivity are specially to be mentioned.

The performance data of single elements as well as interconnected elements were determined in a test rig (9). For the gas/gas high-temperature use the results for the ceramic heat exchanger elements with rectangular channels (4,8x23mm) are indicated. The geometric data of this type of element are shown in table 4.

First of all a single element was tested in pure cross stream. In the second test three elements are connected in the flue gas direction in line, for the air stream they are connected parallelly. For the third test the cross counter stream connection was measured by installing reverse chambers on the air side. With a mass flow of 600 kg/h and a flue gas temperature of 750°C a single element achieves appr. 250°C air temperature in a pure cross stream (fig. 10). Three elements in a line for the flue gas and parallelly for the air rise the temperature to 350°C, with an effective cross stream connection appr. 420°C are reached. Inreasing the flue gas temperature up to 1200°C appr. 380°C, 610°C as well as 790°C are reached. This means an improvement of 88%.

The thermal efficiency is shown as the relation of maximum transmittable power to the really transmitted power (fig. 11). For 600kg/h mass flow and 1200°C flue gas temperature the thermal efficiency for the single element is 28%, for line-parallel-connection 48% and for the cross counter flow 68%. The reduction of the exhaust gas temperature leads to a clearly decreased efficiency.

The necessary blower power can be derived from the data for the pressure losses (fig. 12). They diminish the total efficiency for a certain percentage. However, it can be realized out of this example that the losses are not too high. Naturally, the pressure losses increase with the total length to be flowed through and the higher mass stream. In the cross stream connection, the free flowed through air section is three times higher as at the two others. So the ratio between pressure drops in the flue gas to the pressure drop in air is reversed. The largest pressure drop occurs at the real cross counter stream connection and amounts at 600 kg/h to appr. 1600 Pa for the air side and to appr. 850 Pa for the flue gas side.

In a gas/liquid connection two standard cross stream modules were connected to a cube with 300 mm edge length and put to the flue gas stream. The flue gas is cooled with water by outer reverse chambers in a tenfold cross counter stream (fig. 13). An exhaust gas temperature between 250°C and 1200°C and a mass flow from 270 kg/h to 600 kg/h was used. The cooling water stream was limited to 2000 kg/h and had an inlet temperature to the heat exchanger of constantly 12°C.

At a 250°C flue gas temperature and a low mass flow the condensing field is reached. One can conclude from those results that this unit can bring out appr. 200 kW power from a flue gas stream. During all tests this type of heat exchanger always remained at the temperature of the cooling water.

5. CONNECTING TECHNIQUE

The maximum size of the elements for the ceramic heat exchanger is limited due to the production techniques. Normally it is impossible to heat-seal or solder ceramic parts. This problem was solved by the module concept and a ceramic adapted joining and connecting technique. The open block shape of the heat exchanger elements with always the same outer dimensions, but variable inner structure permits a module construction of larger units with very different arrangements for industrial use.

Joining and sealing of the elements against each other and to periphery is made in the direction of one product stream. In principle, the same techniques as they are used for constructing graphite heat exchangers can be applied. The real sealing element for high temperatures for example can be a ceramic fibrous felt. For the use of liquid and abrasive media graphite and PTFE seals as well as glass solder combinations can be used. Thereby the block shape offers sufficient space for reliable sealings between the blocks and for the joining elements.

6. CERAMIC HEAT EXCHANGERS IN INDUSTRIAL USE

Since December 1986 in different porcelain kilns up to 6 single elements are in service. Due to the high flue gas temperature of 1430°C and the corrosive substances that escape from porcelain glaze no metal heat exchangers can be used. The elements are used in the cross stream system, each of them provides preheated combustion air of appr. 440°C for two burners. Due to recuperation appr. 25% energy will be saved from a complete kiln run.

The example in fig. 14 shows two heat exchanger units with each two elements in cross counter stream connection. The modules which are parallelly connected in the flue gas stream (1200°C) of a high temperature tunnel kiln preheat combustion air for the kiln to appr. 700°C.

In a movable heat exchanger unit (fig. 15) three elements are used. The flue gas flows linear from below to above through the three elements, the air is conducted through the three elements in the cross counter stream system. The compact ceramic unit has a heat exchanging area of 4,35m² with outer dimensions of 100x700x600 mm and a weight of 280 kg. With this unit combustion air temperatures of 1100°C were already achieved. A kiln manufacturing company already produced different apparatus with two respectively three elements.

A unit containing 15 ceramic heat exchanger elements (fig. 16) was built for cleaning air. The first level with 9 elements preheats gas in 3 parallel tractions to appr. 500°C. A second level with 6 elements uses the remaining energy of the exhaust gas to produce 200°C hot air for a dryer.

An example for chemical appliances is the acid cooler in fig. 17. This acid cooler is used for cooling concentrated sulphuric acid having a temperature of 250°C. Neither metals nor graphite can be used in such a case. Six elements are conncected in a line on the acid side and are cooled by water or vapor lying in an enamelled steel container. The feeding pipes are also lined with corrosion resistant materials.

7. CONDENSING HEATING UNIT FOR DOMESTIC APPLICATION

The domestic boiler the schematic view of which is shown in fig 18 has been developped in cooperation with the Institut für Reaktorentwicklung of the KFA in Jülich, RFA. Besides the usual peripheric equipment of a boiler it consists of a gas blow burner (6), a water cooled metallic combustion chamber (7), a ceramic heat exchanger (9) and condensation separator (10).

The device can be used in switch on / switch off service, but can also work modulating. The hot gas is conducted from above to below and cooled down by the heating water circulation from firstly 1000°C to an exhaust temperature which is appr. 5°C above the heating water back flow temperature. The heating water is conducted in counter flow and extracts appr. 60% of the energy of the hot gas in the ceramic heat exchanger, the remaining rest is absorbed in the metallic combustion chamber. The combination of an additional heat exchanger and a range boiler enables preheating of water for domestic use.

Fig. 19 shows the heat exchanger without lining in side view. In table 5 the important performance data of this heat exchanger are indicated. With these data the requirements of the German Standard according to DIN 4702/part 6 are fulfilled.

8. SUMMARY

* A modular constructed ceramic heat exchanger system has been developed. The component is an open block element with rectangular or round channels.

* The block structure can be adapted to special requirements with the aid of flexible production in tape cast technique, for example in the use of particlefilled flue gas streams or condensation service.

* Serveral heat exchanger units were installed and tested in different processes in the last two years.

* By use of a ceramic counter flow heat exchanger a condensing boiler has been developed. Related to the net calorific
value in condensing service the efficiency is 104%. After a
test cycle of 3 years no corrosion can be observed in the
heat exchanger.

9. REFERENCES
1. B.D. Foster, J.W. Patton: Advances in Ceramics. Vol. 14,
   Am. Ceram. Soc. Inc., Columbus, Ohio, 1985

2. J.C. Williams: Doctor-Blade Process. In: F.F.Y. Wang:
   Treatise on Materials Science and Technology. Vol. 9.
   Academic Press., New York, San Francisco, London, 1976,
   173-197

3. G.Y. Onada, Jr.: The Rheology of Organic Binder
   Solutions. In: G.Y. Onada, Jr., L.L. Hench: Ceramic
   Processing Before Firing, John Wiley and Sons, Inc.,
   New York, Chichester, Brisbane, Toronto, 1978, 236-251

4. R.E. Mistler, D.J. Shanefield, R.B. Bunk: Tape Casting of
   Ceramics. In: G.Y. Onada, Jr., L.L. Hench: Ceramic
   Processing Before Firing. John Wiley and Sons, Inc.
   New York, Chichester, Brisbane, Toronto, 1978, 411-448

5. D.W. Richerson; Modern Ceramic Engineering. Marcel
   Dekker. Inc. New York, Basel, 1982

6. J. Heinrich, H. Schelter, S. Schindler, A. Krauth:
   Process for Manufacturing Heat Exchangers from Ceramic
   Sheets, US-Patent 4.526.635 (1985)

7. R. Ganz, H. Schelter, O. Heinz: Wärmetauschermodul aus
   gebranntem keramischem Material EP 0 274 694

8. S. Förster, P. Quell, J. Heinrich, J. Huber, H. Schelter:
   Ceramic Residential Boiler with Condensation of
   Combustion Water Vapor, Proc. Int. Symp. Condensing Heat
   Exchangers, 1987, Columbus, Ohio

9. J. Heinrich, J. Huber, H. Schelter, R. Ganz, R. Golly, S.
   Förster, P. Quell: Compact Ceramic Heat Exchangers:
   Design, Testing and Fabrication British Ceramic
   Transactions and Journal Vol. 86 (6) 170-205 (1987)

Table 1: Characteristic data of a counter flow heat
         exchanger for a condensing boiler

Characteristics	Unit	Value
volume total	dm^3	4,85
content of water	dm^3	0,87
geometric heat transfer surface	m^2	0,9
hydraulic diameter of flow channels		
.water	mm	3,15
.combustion gas	mm	4,0
weight total	kg	7,6
height, width, depth	mm	258/145/116

Table 2: Corrosion of SiSiC in different media

Medium	Temperature ^{o}C	Weight Loss in mm/year
Sulphuric acid	200	0,018
Hydrochloric acid	180	0,019
Nitric acid	200	0,007
Acetic acid+3%anhydride	Boiling point	0,003
Acetic acid+3%anhydride +100ppm chloride as NaCl	Boiling point	0,005
Formic acid	Boiling point	0,005
Formic acid+100ppm chloride as NaCl	Boiling point	0,004
Hydrofluoric acid	Boiling point	5,49
Sodium hydroxide 50%	150	dissolved
Deionized water	200	0,304
Deionzied water+100ppm chloride as NaCl	200	0,288

Table 3: Properties of tape cast SiSiC

Property	Unit	Value
Density	g/cm^3	3.0
Porosity	%	0
Gas permeability (tape)	$kg/sec \times cm^3$	0
Thermal conductivity at RT	W/mK	120
Max. service temp.		
(oxidizing)	$^\circ C$	1400
(reducing)	$^\circ C$	1400
Spec. thermal capacity ($20..1000^\circ C$)	$J/kg \times K$	950
Coefficient of linear expansion ($20..1000^\circ C$)	$10^{-6} K^{-1}$	4.4
Flexural strenghth	MPa	400
Young's modulus	GPa	370
Alkali resistance		moderate
Acid resistance (except HF)		excellent

Table 4: Geometric characteristics of a cross flow
heat exchanger with rectangular channels

Characteristics	Unit	Value
Outer dimensions	mm	300x150x300
Volume total	cm^3	13500
void fluid 1	cm^3	3809
void fluid 2	cm^3	3809
Mass total	kg	17.1
Hydraulic diameter		
fluid 1	mm	7.95
fluid 2	mm	7.95
Flow path length		
fluid 1	mm	150
fluid 2	mm	300
Geometric heat transfer surface	m^2	1.6
Specific heat transfer surface (related to total volume)	m^2/m^3	118.5
Specific surface weight	kg/m^2	10.69

Table 5: Performance data of the condensing ceramic boiler

Property	Unit	Value
Power range	kW	9 - 19
Pressure drop (total)		
- gas side	mbar	0,6 - 0,7
Pressure drop		
- gas side	mbar	0,1 - 0,2
- water side	mbar	2
Combustion chamber temperature	$^{\circ}C$	1000
Temperature difference between flue gas exhaust and back flow water	$^{\circ}C$	5
Condensation rate of H_2O vapor arising at combustion	%	>80
Efficiency related to the net calorific value	%	104
Flue gas emission within the power range at excess air =1,1		
Co	ppm	33 - 65
Co_2	ppm	10,7
No_x	ppm	40 - 45
pH-Value condensation		2 - 3

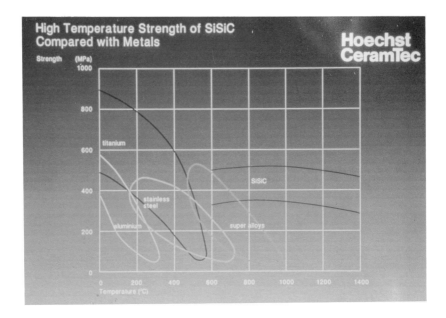

fig. 1: Strength of silicon infiltrated silicon carbide
compared with metals

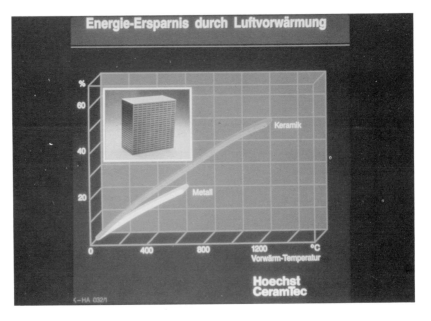

fig. 2: Engergy savings by preheating combustion air

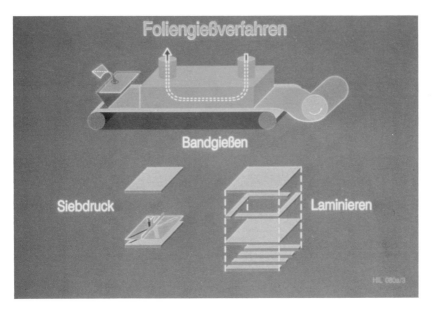

fig. 3: The tape casting technology.

fig. 4: Construction of a ceramic cross flow heat exchanger with rectangular channels by laminating tapes and stripes.

fig. 5: Construction of a ceramic heat exchanger by
         laminating punched cards.

fig. 6: Different heat exchanger designs.

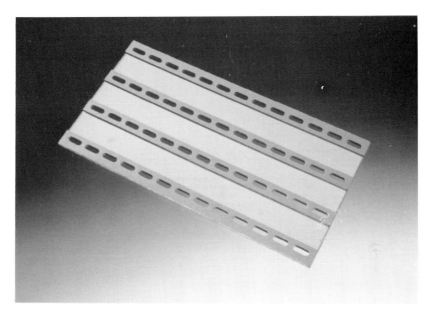

fig. 7: Rippled basis plate for a gas/liquid heat exchanger.

fig. 8: Gas/liquid heat exchanger with integrated connections.

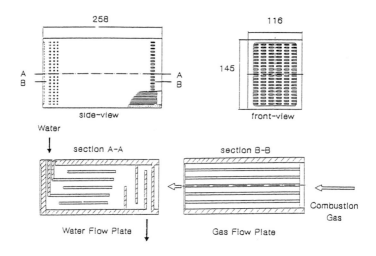

fig. 9: Basis Design of a gas/water counter flow heat
        exchanger.

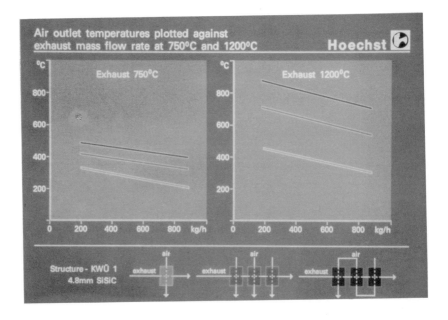

fig. 10: Air temperature vs mass flow at different exhaust
         gas temperatures.

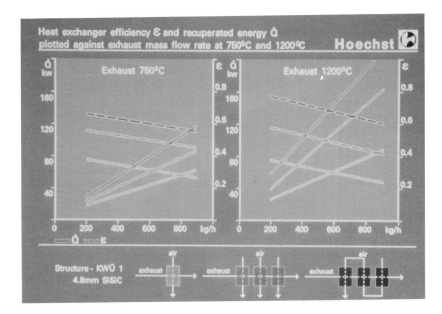

fig. 11: Thermal efficiency and recuperaterd energy vs
mass flow at different exhaust gas temperatures.

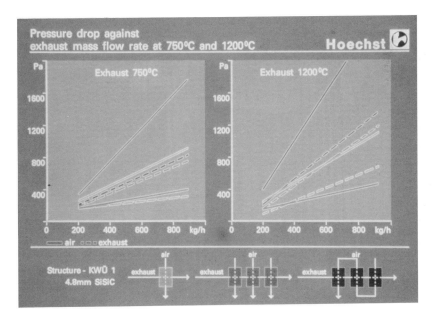

fig. 12: Pressure drop vs mass flow at different exhaust
gas temperatures.

fig. 13: Condensing heat exchanger unit with 2 standard
        elements.

fig. 14: Two recuperator units with each two elements in
        cross counter flow connection.

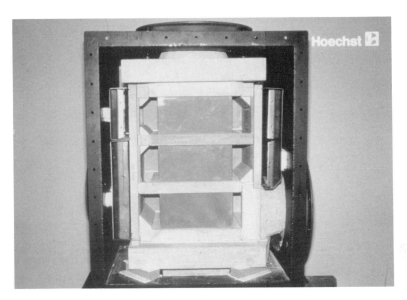

fig. 15: Movable heat exchanger unit with three elements
in cross counter flow connection.

fig. 16: Heat exchanger unit with 15 elemtents for air cleaning.

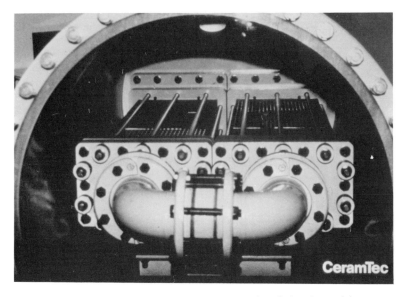

fig. 17: Acid cooler for concentrated sulphuric acid.

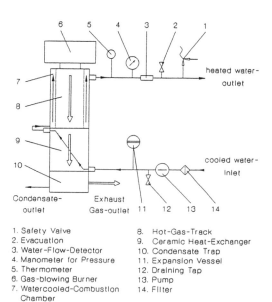

1. Safety Valve	8. Hot-Gas-Track
2. Evacuation	9. Ceramic Heat-Exchanger
3. Water-Flow-Detector	10. Condensate Trap
4. Manometer for Pressure	11. Expansion Vessel
5. Thermometer	12. Draining Tap
6. Gas-blowing Burner	13. Pump
7. Watercooled-Combustion	14. Filter
Chamber	

fig. 18: Schematic view of the condensing ceramic boiler.

fig. 19: Condensing ceramic boiler.

Development of a ceramic high temperature recuperator—economic analysis through prototype unit

G. R. Peterson, P.E.
U. S. Department of Energy, Idaho Falls, ID
and
W. P. Parks, and J. R. Bower
Babcock and Wilcox, Lynchburg, VA

ABSTRACT: Two significant obstacles in the commercial acceptance of ceramic heat exchangers are the high cost compared to conventional metallic units and the difficulty in predicting ceramic performance. An Oak Ridge National Laboratory (ORNL) study examined how future changes in ceramic manufacturing costs will affect ceramic recuperator cost. In a separate project, Babcock and Wilcox (B&W) is developing destructive and nondestructive test methods to enhance ceramic reliability. Finally, in a third project, B&W designed, installed and tested an innovative ceramic High Temperature Burner Duct Recuperator (HTBDR).

INTRODUCTION

The Advanced Heat Exchangers Program (AHX) of the U.S. Department of Energy (DOE) Office of Industrial Programs (OIP) sponsors technology development to advance the state-of-the-art in heat recovery and to expand U.S. energy conservation. High-temperature flue gases are an untapped energy source in many plants. To encourage the recycling of this waste energy, OIP created an integrated program to overcome the major obstacles to high-temperature waste heat recuperation. This paper discusses the program response to three of these commercial impediments: the high cost of ceramics compared to metals, the difficulty in predicting ceramic performance, and the dearth of applied research in high temperature ceramic recuperators.

CERAMIC RECUPERATORS - COST FACTORS

A recuperator is a gas-to-gas heat exchanger that recaptures waste heat. The industrial demand for ceramic recuperators remains sluggish due to cost factors, despite proven technical performance. In an industrial furnace ceramic recuperators may diminish fuel consumption by up to 50% with a 1100C combustion air preheat. By comparison, a typical metallic recuperator will reduce the fuel consumption of an industrial furnace about 20-30%. But despite the higher thermal efficiency, the high initial capital costs of the ceramic recuperator renders these units economically unattractive at current natural gas prices (Parks and DeBellis).

The economic obstacle course facing ceramic recuperators can be illustrated by the B&W HTBDR, developed under DOE sponsorship. This ceramic heat exchanger is a bayonet, i.e., a tube-within-a-tube design, with 50 SiC tubes suspended from an insulated tubesheet. This recuperator was successfully tested in 1200C flue gas from a steel soaking pit, producing 750C preheated combustion air. B&W estimates fuel savings of 17% compared to a single metallic heat exchanger (Parks and DeBellis).

Fuel savings are estimated at 41% compared to an unrecuperated furnace. The technical aspects of this HTBDR will be discussed later in this paper. However, despite the demonstrated successful technical performance, the B&W HTBDR has yet to succeed commercially directly because of ceramic cost. B&W estimates the installed commercial cost of this ceramic HTBDR unit at $314,000. Cost would be optimized relative to performance for any given application. Under the test operating conditions the cost of the energy recovered from the ceramic recuperator would be about $7/MMBTU. This does not compare favorably with current natural gas prices in the U. S. of about $3/MMBTU.

The cost of ceramic heat exchangers will decline in the future as the cost of ceramic processing decreases. Therefore, ORNL, under OIP sponsorship, studied the effect that decreasing ceramic production costs would have on the cost of a ceramic recuperator (Das et al 1988). Currently, a 140 m^2 ceramic recuperator costs about 70% more than a comparable metal unit. Furthermore, a metal recuperator with ceramic tubes costs about 50% more than a comparable all metal heat exchanger. However, improvements in ceramic manufacturing technology and economics of scale will tend to drive down the ceramic recuperator cost.

ORNL evaluated two types of ceramic production, slip casting and extrusion. Production costs were calculated from models developed by the Materials Systems Laboratory (MSL) at the Massachusetts Institute of Technology (MIT) and updated by ORNL. The models analyzed six process stages in the production of a SiC heat exchanger tube: a) SiC powder preparation, b) extrusion/casting, c) drying, d) firing, e) machining, and f) quality control and storage. The model factored into each process stage, five cost categories: a) materials, b) energy, c) labor, d) capital, and e) overhead. Calculations were based on the production of 5.1 cm OD x 4.1 cm ID x 122 cm long SiC tubes with SiC heat exchanger headers. The model treated ceramic headers as a specialized tube type when estimating costs. Generally, ORNL believes that the following conclusions apply to SiC headers as well as SiC tubes.

Economics of scale obviously influence the cost of SiC ceramic heat exchangers. ORNL has concluded that manufacturing cost decreases exponentially and reaches an asymptote at production volumes of about 20,000 tubes/yr. When production levels are doubled from 10,000 tubes/ year to 20,000 tubes/year, ORNL estimates that cost will decrease from $550/tube to $150/tube. A further doubling of the production volume, 40,000 tubes/year, reduces the cost by about only $20 to $130/tube. The cost of slip casting as a function of production volume is comparable to that of extrusion.

Tube cost is a strong linear function of the ceramic powder cost because materials represent approximately 60% of the total tube cost. ORNL estimates that each $22/kg decrease in the price of SiC powder will reduce cost by about $70 per tube.

Quality Assurance in ceramics manufacturing can significantly decrease the number of nonconforming ceramic tubes and hence dramatically decrease tube cost. ORNL estimates that reducing the amount of scrap and rework by 50% will decrease SiC tube price by about 20%. Because product quality is critical to industrial acceptance of ceramic recuperators, B&W, supported by the INEL Research Center (IRC), is currently developing destructive and nondestructive test techniques to determine and detect the strength limiting flaws of SiC ceramic heat exchanger tubes. This project is discussed below.

Although at present ceramic recuperators cannot compete directly with their metal counterparts, the ORNL study suggests that a change in the ceramic manufacturing technology may produce more cost competitive ceramic recuperators. Improved manufacturing techniques will reduce costs which will increase production volume which will, in turn, reduce costs. The areas promising the greatest cost benefits are SiC powder cost reduction and improvement in ceramic quality assurance. ORNL analyzed SiC tube cost as a function of the energy, capital and labor costs and concluded that tube cost is not particularly sensitive to anticipated variations in these parameters (Das et al 1988).

CERAMIC RECUPERATORS - STRENGTH LIMITING FLAWS

Industrial acceptance of ceramic recuperators is hindered by the unreliability of the ceramic components. As noted above, improved quality assurance techniques, such as the measurement of the critical flaw size, can substantially reduce the final production cost. Furthermore, measuring the critical flaw size and understanding its behavior will enable a ceramics manufacturer to predict tube lifetime and hence provide a warranty. Conversely, based on a desired lifetime and specified operating conditions, the manufacturer will be able to determine the maximum allowable flaw size in the ceramic material. Babcock and Wilcox (B&W), supported by the IRC, is currently researching the strength limiting flaws of SiC heat exchanger tubes. Research is progressing along two paths: a fracture mechanics model to predict ceramic tube lifetime, and test methodology to measure the size of the strength limiting flaws.

This project assumes that linear elastic fracture mechanics (LEFM) concepts predict ceramic brittle fracture behavior. LEFM compares the stress intensity factor of the critical flaw with the material's resistance to crack growth. For example, for some C-ring specimens, the Gross and Srawley equation may be used:

$$K_I = M \times \sigma \times \sqrt{(\pi a/Q)} \quad (1)$$

$$Q = 1 + 1.464(a/c)^{1.65} \quad (2)$$

where: K_I = mode I stress intensity factor
M = empirical parameter dependent on crack geometry
σ = applied stress
a = maximum depth of surface flaw
$2c$ = flaw length at free surface
Q = a geometry correction term based on a and c

The Gross and Srawley equation is a relationship only for C-ring samples with edge notches on the outside diameter. Furthermore, the notch must be within a certain range of geometric parameters. Analogous equations exist for various specimen geometries and for internal flaws.

K_{IC}, the plane strain fracture toughness, expresses the magnitude of the elastic stress field at the crack tip at the instant of fast fracture. K_{IC} is an intrinsic property of the material and accordingly independent of the flaw size or geometry. Given K_{IC} and M, the relationship between the design stress, σ, and the maximum allowable flaw, a and c, is established.   Note that the maximum initial flaw is inversely proportional to the observed failure stress.

K_{IC} is the stress intensity at fast fracture. However, there are two types of crack growth, slow and fast. Slow crack growth may persist for extremely long times. At some point, a flaw may grow until it causes fast fracture, a process called static fatigue. In addition, the strength degradation at temperatures >1000C may involve flaw nucleation and coalescence as well as flaw growth. A stress intensity factor, K_{th} is assumed where $K_{th} < K_{IC}$. K_{th} is the threshold value of K where the critical flaw starts to grow prior to fast fracture. If K_{th} can be conservatively calculated for a given set of industrial conditions, then a time to failure of the ceramic tube can be estimated. Surface flaws are assumed to be of more concern than interior flaws. The value of M is usually higher for a surface crack as opposed to a similar flaw in the interior. Furthermore, a surface crack is susceptible to growth through corrosion. For example most nitrides and carbides tend to oxidize at temperatures in excess of 1000C. Therefore, surface defects most often initiate failure in ceramic materials.

Babcock & Wilcox, (B&W), and the IRC, are currently creating the methodology to nondestructively measure ceramic strength limiting flaws. B&W and IRC use a combination of Time-of-Flight Acoustic Microscopy (TOFAM) and Computed Tomography (CT) X-rays. Both NDE techniques possess a lower detection limit theoretically equal to the grain size of the SiC sample. The TOFAM is selected to detect sharp cracks while the CT system is better for the detection of ceramic inclusions. The TOFAM scans the SiC sample with ultrasonic energy and times the echo return. Computed Tomography (CT) was first developed for use in medicine but its potential NDE applications are just now beginning to be realized. CT provides a high resolution image of a sample slice with high contrast sensitivity. B&W and IRC are currently refining the TOFAM and CT systems to improve analysis speed and reliability.

The destructive testing path of this project attempts to define the relationship between flaw size/shape and the expected industrial SiC tube lifetime. The test matrix contained both C-rings and tube sections, both natural and seeded defects, and both thermally-cycled and isothermal ambient temperature samples. After fracture, the samples were analyzed by Scanning Electron Microscope, (SEM), to measure the critical flaw. Features such as surface roughness, coarse hackle markings, and hinges were noted. Currently the project team can determine the SiC sample fracture toughness, K_{IC}, and predict the stress that produces brittle failure. B&W and IRC are continuing research to predict the minimum SiC tube lifetime in a high temperature, approximately 1200C, air environment based on the slow crack growth rate.

Based on the calculated K_{IC}, B&W and IRC predicted the pressure at SiC tube failure at a given flaw size and then conducted pressurized tube tests for confirmation. Test results are incomplete, but, except for one test, the tubes burst at a pressures roughly 30% higher than that predicted. This suggests that the B&W/IRC test methodology could serve as a conservative design tool.

Acoustic Emission (AE) techniques qualitatively detect slow crack growth by detecting the sound emitted when a grain breaks or a grain boundary opens. AE events were detected in the pressurized tube samples. Tube samples were stressed in increments of 1 MPa (150psia) until failure. AE sensors located around the tube circumference detected the magnitude and arrival time of each AE event. Back projection was then used to calculate the event location and activity. The AE events seem to correlate well with the location of the fracture planes as determined by post test fractography. AE events, possibly indicating slow crack growth, start at about 30% of the breaking load. B&W and IRC are continuing research to relate the AE data to K_{th} and slow crack growth.

Both B&W and IRC are continuing this project. Although the initial data are encouraging, the preliminary results are incomplete and, therefore, not presented in this paper. The NDE methodology may be able to detect the allowable flaw size if it is significantly larger than the background grain size and surface roughness. Given the critical flaw size, fracture mechanics can then predict the stress that produces brittle failure in a SiC tube. The research objective now is to predict the minimum lifetime of a SiC tube in a high temperature, approximately 1200C, air environment. The minimum lifetime prediction requires quantification of the slow crack growth rate and hence a reliable prediction of K_{th}. The project will attempt to relate K_{th} with the AE rates and determine K_{th} as a function of temperature and induced stresses. Future work also includes refining the TOFAM and CT systems to improve analysis speed and reliability.

CERAMIC RECUPERATORS - PROTOTYPE DEVELOPMENT

A recuperator is a gas-to-gas heat exchanger that recaptures waste heat. Many plants, particularly in the steel industry, exhaust high temperature flue gas. Metallic recuperators recover waste heat from these types industrial processes but are generally limited to flue gasses <900C. For higher temperature flue gasses, diluent air may be used to lower the flue gas temperature at a cost of thermal efficiency. Therefore B&W, in a cost-shared program with DOE, developed a High Temperature Burner Duct Recuperator (HTBDR) based on SiC and designed to survive in oxidizing environments up to 1250C. This particular unit preheated the combustion air in a steel mill soaking pit to 750C thereby reducing furnace fuel consumption. The ceramic HTBDR secured an additional 17% fuel savings over the previous recuperated system, a metallic recuperator. Fuel savings would be at 41% compared to an unrecuperated furnace (Parks and DeBellis).

The ceramic heat exchanger uses a tube-in-tube bayonet concept as illustrated in Figures 1 and 2 (Parks and DeBellis). The bayonet concept does not constrain the lower end of the tube thereby reducing the stresses. Furthermore, the heat exchanger was designed to take maximum advantage of radiation heat transfer, minimize pressure drops on both air and flue gas sides, and minimize thermal stresses. The ceramic heat exchanger consists of fifty outer tubes of siliconized SiC, each containing an inner tube. The inner tubes are recrystallized SiC on the upstream end and metallic on the downstream end. The tubes are supported by two air-cooled metallic tubesheets with Koawool fiberglass insulation. Sleeves connect the tubes with the tubesheet and are resilient enough to adjust for thermal expansion. The seals prevent air movement between combustion air and flue gas streams. B&W selected SiC tubes because SiC possesses good thermal shock resistance, strength, thermal stability and low permeability. SiC coupon testing at the host site confirmed material suitability. Ceramic fiber materials were chosen for the sleeve/seals.

Seals of ceramic filter paper comply with moderate thermal expansion while
bonded ceramic fiber board sleeves provide the required rigid support.

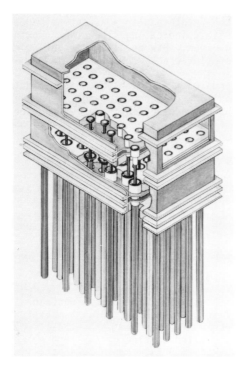

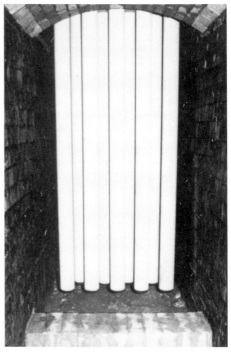

Fig. 1 Ceramic Stage   Fig. 2 Ceramic Stage Installed

The HTBDR was installed in a steel mill soaking pit at a B&W plant near
Beaver Falls, Pennsylvania. The HTBDR system is illustrated in Figure 3
(Parks and DeBellis).   In operation, the combustion air initially passes
through a metallic heat exchanger, is heated to 430C, and then passes to
the ceramic heat exchanger.   The air moves into the upper plenum, down
through the inner tubes, up through the space between the tubes and exits
through the lower plenum.   The combustion air leaves the ceramic heat
exchanger at a temperature of 750C. The preheated air then moves through
the insulated ducts to the burner. Steel ingots are heated in the pit by
the hot combustion gasses. Spent flue gas exits the pit and reaches the
ceramic heat exchanger at temperatures up to 1200C. The ceramic heat
exchanger cools the flue gas to a temperature of around 875C where it can
then pass on to the metallic heat exchanger.

The field test verified that the prototype HTBDR could successfully
operate for at least 1400 hours.   Table 1 gives the data (Parks and
DeBillis).   No tube failures occurred. Several leak tests verified no
detectable leakage through the seals. After the test, B&W examined the
heat exchanger tubes and did not observe any significant tube recession or
corrosion although there was some surface reaction as is to be expected in
SiC.   Despite the successful technical performance, industry has yet to
accept the recuperator.   Current energy prices are subverting B&W's

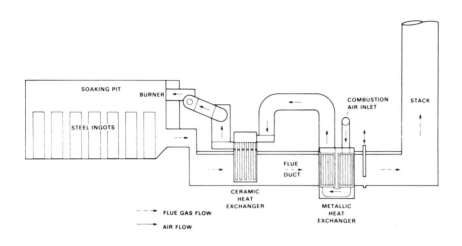

Fig. 3. Schematic of HTBDR System as Installed

Table 1. HTBDR Field Test Data

Flow Rates, kg/hr	
Natural Gas	84.8
Combustion Air	2830
Flue Gas	2330
Temperature,C	
Ceramic Stage	
Air In	435
Air Out	745
Flue Gas In	1190
Flue Gas Out	875
Metallic Stage	
Air In	38
Air Out	435
Flue Gas In	840
Flue Gas Out	445

Table 2. HTBDR Installed Costs

System	Ceramic	Metallic	Both
Burner	$ 15,000	$ 7,000	$ 15,000
Ducting	$ 20,000	$10,000	$ 20,000
Fan	$ 15,000	$ 7,500	$ 15,000
Control	$ 30,000	$15,000	$ 30,000
Install	$ 40,000	$30,000	$ 45,000
Met. HX	$ ------	$25,000	$ 25,000
Cer. HX	$194,000	$------	$194,000
Total	$314,000	$94,500	$344,000

calc based on 1987 dollars

commercialization efforts. The proper comparison in an economic analysis of the HTBDR is the ceramic heat exchanger cost versus the conventional metallic heat exchanger cost. Although the B&W HTBDR system delivers impressive energy savings, about 60% of these energy savings are attributable to the metallic recuperator which only accounts for about 30% of the cost.

At the host site steel soaking pit, the HTBDR system saves about 6.5 x 10^6 KJ/hr under normal operating conditions. This is the equivalent of about \$22/hr assuming a natural gas price of \$11.80/100m^3. A typical production run averages about 20 hours, of which the burners run for 7 hours at high fire and at the remaining 13 hours at low fire. Under high fire, the HTBDR system conserves about 12 x 10^6 KJ/hr while under low fire conditions the recuperator system saves about 3 x 10^6 KJ/hr.

Table 2 presents the B&W estimate for the total installed costs of the HTBDR system. The metal recuperator represents about 30% of the total cost. By far, the most expensive component is the ceramic heat exchanger at \$194,000. Furthermore, the ceramic tubes represent the greatest cost to the ceramic recuperator. For example, one SiC outer bayonet tube costs about \$1,400.

As mentioned above, the proper comparison in an economic analysis is the ceramic heat exchanger versus the conventional metal heat exchanger in performance, cost, maintenance downtime, and other factors. For the prototype HTBDR system described, not optimized, assuming 890C preheat air temperature, 6240 hr/yr operation and a natural gas price of \$12/100m^3, the annual fuel savings from the ceramic unit alone equals \$37,000. This translates into a simple payback of about 9 years (Parks and DeBellis). These simple payback calculations are based on the estimated costs for the first commercial HTBDR. Subsequent units should cost less.

SUMMARY

The B&W HTBDR has found slow acceptance in the market-place because of the high ceramic component costs and low natural gas prices. However, increased ceramic production, improvements in ceramic manufacturing, and the introduction of ceramic NDE techniques will, in the future, reduce ceramic prices. Furthermore, it is reasonable to assume that energy prices will increase in the future. Therefore, the expanding use of ceramic heat exchangers for energy recuperation and a host of other industrial processes may be just a matter of time.

ACKNOWLEDGMENTS

The authors wish to thank T. Randall Curlee, Oak Ridge National Laboratory, and W. R. Lloyd, IRC, for invaluable advice in the preparation of this paper. Appreciation is also extended to S. L. Richlen of U.S. DOE for guidance during the projects described in this paper.

REFERENCES

Das, S., Curlee, T. R. , and Whitaker, R. A., *Ceramic Heat Exchangers: Cost Estimates Using a Process-Cost Approach*, Final Report, ORNL, ORNL/TM-10684, August 1988.
Parks, W. P., and DeBellis, C. L., *High Temperature Burner Duct Recuperator System Evaluation*, Final Report, B & W, DOE 12296-5, (to be published)

Paper presented at Conf. on Ceramics in Energy Applications, Sheffield, April 1990
Session 5

Some aspects of energy conservation at British Steel Stainless

P E Davis M.Inst.Ref.Eng. and G Payne F.I. Ceram.

ABSTRACT: Energy conservation has always been a consideration in the design of heating furnaces, but it has become of increasing importance in recent years and encouraged the exploitation of new technologies. British Steel Stainless has been particularly active in this field and this paper describes some of the development work of interest. Furnace design and operational aspects are discussed, but the main emphasis is on the role of refractories.

1.   INTRODUCTION

British Steel Stainless, a division of British Steel plc, is one
of the world's largest stainless steelmakers. Approximately 2,400
people are employed in the Sheffield area and a further 547 in South
Wales.   Hot rolled plate and cold rolled material are produced in
Sheffield with the Panteg Works, in South Wales, producing cold rolled
material and long products.   The Sheffield (SMACC) plant produces in
excess of 350,000t of liquid steel/annum via a 130t electric arc furnace
feeding a 130t AOD vessel.   A recently installed ladle furnace is now
in operation and a plasma smelting furnace, designed to recover valuable
metallic elements from dust collected by the fume extraction plant,
is a further recent addition.   The steel is continuously cast in a
slab casting machine producing slabs 950-1550mm wide by 150-200mm in
thickness.

Plate is produced on a four-high, 2m reversing mill working in
conjunction with finishing facilities capable of producing plate up
to 3m wide.

Softening and descaling lines and a vertical bright annealing furnace
are available for the production of stainless strip.

Within this area the main energy usage is concerned with the heating
of steel prior to rolling, heat treatment of rolled material and other
activities such as the heating of melting shop ladles.   The heating
and heat treatment is carried out in 10 assorted furnaces so that there
is considerable incentive to develop thermally efficient units.   This
has long been recognised and much work has been done.   This paper
describes some of this work, but concentrates particularly on the role
of refractories.

2.   TECHNICAL DEVELOPMENTS

The Iron and Steel Industry is one of the largest individual consumers
of energy in the UK, accounting for around 5% of the total energy
consumed in the country.   Within British Steel, energy requirements
represent approximately 15% of the total manufacturing costs.   Stainless
steel products in particular demand a high energy input and in recent
years considerable emphasis has been placed on exploiting new energy
saving techniques.   The results have been dramatic.   Over the last
decade, technological developments together with production changes
have resulted in a 56% reduction in the amount of energy required to
produce a tonne of finished stainless steel (Figure 1).   This improvement
is the result of a sustained development programme, involving both
works and technical staff and has been a major factor in the
profitability of the business.

Mr P E Davis, British Steel Stainless, Technical Services Department
Mr G Payne,   British Steel Technical, Teesside Laboratories

In broad terms, there have been four main areas of development.

1. The use of more thermally efficient burners.

2. Changes in furnace design and geometry.

3. The use of more thermally efficient refractories, particularly ceramic fibres.

4. Exploitation of operational benefits resulting from these changes.

The application of self recuperative and more recently, regenerative ceramic burners have been described elsewhere in some detail[1] and it is not proposed, therefore, to discuss this topic further in this paper; it should be emphasised, however, that the contribution made by installing the more efficient burners has been considerable.

Brief reference will be made to furnace design changes and operational benefits will also be discussed, but the main emphasis in this paper will be to consider the role of refractories in energy conservation.

3. FURNACE DESIGN AND GEOMETRY

In order to fully exploit new burner technology, it was essential to modify existing furnace geometry. Changes to the shape of the furnace interior can improve the movement of the hot gases and assist heat transfer from the furnace lining to the stock. Sophisticated modelling techniques are available which provide a rapid and relatively low cost method of determining the optimum furnace configuration. This part of the work was carried out by the Fuel and Furnaces Department, British Steel Technical, Swinden Laboratories, based in Rotherham and an account of some of this work is given elsewhere[2].

In order to obtain good temperature control and combustion efficiency, each furnace needs to be considered individually. Savings will depend on establishing the correct thermal input and burner configuration but, in addition, good stock temperature distribution and furnace output will depend on achieving the correct furnace geometry.

The decision to carry out these major changes provided the opportunity to consider in some detail the refractory requirements. The large, high heat content structures, typical of many mill furnaces were very inefficient, particularly when used for batch processing and there was a clear incentive to develop more thermally efficient but also durable, low thermal mass linings.

Figure 2 shows an example of one of these early furnaces and Figure 3 illustrates the significant design changes which were made. The dense fireclay bricks used in the upper sidewalls and roof were replaced by ceramic fibre. Well insulated monolithic refractories formed the lower parts of the furnace.

4.   REFRACTORY DEVELOPMENTS

A valuable contribution has been made to the energy conservation programme by exploiting the unique properties of refractory fibres. These materials are characterised by:

1.   Low thermal conductivity.

2.   Low heat storage

3.   Low density

4.   Good thermal shock resistance

5.   Good retained resilience

6.   Good chemical resistance.

The combination of good insulating properties and low thermal mass has enabled significant benefits to be achieved over a wide range of heating applications.   Table 1 illustrates the relative thermal performances of a series of alternative roof constructions.   A 300mm thick refractory brick structure has been used as a basis for comparison. In terms of steady state heat loss such a structure has a relatively low thermal resistance and is inferior to more recent monolithic installations.

Ceramic fibre applied as a 50mm thick veneer, as in lining B, will initially reduce the heat loss rate by more than 50%, although the thermal efficiency of the fibre may subsequently deteriorate.   Recent experience, however, would now lead us to recommend the use of 75mm thick veneers for operations above 1000°C.

The insulated mouldable refractory lining C and the full thickness semi-insulated castable lining D, are seen to have similar relative heat loss rates.   Table 1 indicates that when compared to the performance of the uninsulated fireclay brick lining A, monolithic linings such as these can yield savings in heat loss of approximately 60%.   The application of a ceramic fibre veneer will result in further gains, as illustrated in the case of lining E.

Where ceramic fibre is used as a complete lining of sufficient thickness, a low heat loss rate will be achieved.

Of probably more significance is a comparison of the thermal capacity of alternative linings.   On each occasion a furnace is cooled, either as part of a designed operating cycle, or for periodic shut-down, much of the heat stored in the lining is lost.   A low thermal capacity lining will clearly reduce such heat losses and also permit savings by achieving rapid cooling and heating.   Constructions based on semi-insulating castables have been used successfully but, where appropriate, ceramic fibre has a large advantage in this respect and has been widely used.

For many years the choice of fibre was limited to standard alumino-silicate fibres having a claimed maximum service temperature of around 1260°C.   The development of improved high duty materials containing additions of refractory oxides, together with alumina, mullite and mixed fibres has extended the useful application of these materials to significantly higher operating temperatures and has added considerably to the diversity and complexity of this range of refractories as shown in Table 2.

Conditions must be favourable for these benefits to be achieved. Much of the early activity was initiated by the suppliers and installers promoting relatively untried systems, using materials not always suitable for the operating conditions encountered. Costly failures occurred which clearly indicated the need for a wider understanding of ceramic fibre properties and behaviour in order to identify the range and limitations of the products and systems available. The need remains, therefore, for care in the selection and application of these materials.

Of particular importance is the fact that all fibres exhibit shrinkage on heating and this can reach serious proportions well below the manufacturers' recommended service limits. Some measures can be taken to reduce the amount of shrinkage, such as the use of higher quality materials, increasing the density of the fibre by applying compression during manufacture and installation and also using fibre in the stacked bonded form, where blanket material is cut into strips and arranged in an 'edge on' configuration. All of these methods have been used at British Steel Stainless with varying degrees of success. Some indication of the potential and also the limitations of these materials will become apparent in considering the following examples.

There are currently ten furnaces operating in the Stainless area which contain significant amounts of ceramic fibre. It is important to stress that all these furnaces operate on natural gas. This is a very significant feature when considering the performance of fibre, as the use of other types of fuel can lead to rapid deterioration.

The furnaces operate within the range 700-1350°C and as the furnace operating temperature is a critical factor with regard to the performance and durability of these materials, it is convenient to consider their application within the following three temperature ranges.

Low temperatures i.e. <1000°C

Intermediate temperatures i.e. 1000-1200°C

High temperatures i.e. up to 1350°C.

At temperatures up to approximately 1000°C, standard fibre blanket linings have proved highly successful. Typically, these linings consist of a backing lining of mineral wool, used in conjunction with three or four layers of standard blanket material. At these low temperatures we have also used, successfully, two layers of 12.5mm thick blanket and 75mm of mineral wool insulation as a backing lining.

The thickness and design will depend on the external temperature and acceptable heat loss rate. Some form of metallic stud is used for anchorage which can be fitted with a ceramic tip for use at higher temperatures, if required. Figure 4 shows one example of this type of lining and illustrates the large scale of some of these installations. This large, low temperature furnace has a total hearth area of $8m^2$ and can accommodate a maximum job height of 4.5m. The maximum operating temperature of the furnace is 700°C.

Benefits of these linings are the speed and simplicity of construction, the light-weight and low cost. On the other hand ceramic fibre, particularly when used in blanket form, is prone to high shrinkage and has a low resistance to erosion by high velocity gases and abrasive particles.

With these batch type furnaces operating at relatively low temperatures, the reduced heat storage is the main attraction. Calculations have indicated reductions of up to 40% in heat lost to the lining, representing a saving of around 10% in fuel used.

As service temperatures exceed 1000°C the use of blanket fibre becomes more problematical. There are a large number of furnaces operating in the Stainless Mills area at these intermediate temperatures and, as described earlier, considerable development work has been carried out on furnace profiles, burners and lining configurations. In early trials it was quickly established that fibre claimed to operate continuously at 1260°C would not perform satisfactorily in blanket form at temperatures as low as 1050°C. Even high duty materials having a claimed service limit of 1400°C were inadequate. Problems due to excessive shrinkage quickly developed with serious consequences (Figure 5). These experiences have lead us to use the following general guide.

High duty blanket is necessary when temperatures approach 1000°C.

Modules, based on high duty fibres, are necessary above this temperature.

Currently, therefore, these furnace linings consist of fibre installed in modular form and usually incorporate a blanket safety lining positioned on the shell (Figure 6). Safety lining thicknesses have varied. Thicknesses up to 100mm have been used or, conversely, the safety lining has been omitted entirely. From our experiences a 25mm thick blanket has given the best results. The modules are typically 200mm in thickness and consist of high duty fibre used in conjunction with a properly designed metallic anchor system. Fibre quality is not now the main problem. The module design, particularly in relation to the type and disposition of the metallic anchor components, is the most important factor. With regard to anchor quality we would always recommend the use of Type 310 stainless, or even Inconel in these systems. Some estimate of the thermal gradient through the lining is, therefore, essential. This information should be used to determine the anchor position and also the steel quality required. Figure 7 illustrates the type of problem which can occur when, as the lining becomes unstable, gas tracking occurs leading to overheating of the anchorage and subsequent failure.

At operating temperatures above 1260°C, experiences at British Steel Stainless indicate that the selection of manufacturer and particularly the design and installed quality of the system, become much more critical. Lining costs escalate rapidly with these high quality systems and some engineering judgement is required in selection.

For example, typical cost ratios when compared to a simple blanket lining are approximately as follows.

Blanket lining for use up to 1000°C   - 1
Modular lining for use up to 1200°C   - 2
High reliability lining for use above 1200°C   - 4.5

For a typical large mill furnace of the type described in this paper, this means that ceramic fibre lining costs well in excess of £30,000 can be expected.

Under favourable conditions, these materials can perform well at high operating temperatures where reductions in heat losses yield considerable savings in energy. There is, however, a growing awareness of their limitations, particularly when used in hard driven, oil fired furnaces, where performances have been generally disappointing.

Excessive shrinkage is the main problem. When this occurs, remedial measures are limited. Stemming can be applied and trials with ceramic fibre spray materials have shown some promise, but excessive shrinkage usually means complete replacement of the lining.

The behaviour of ceramic fibre at these high temperatures is complex and is dependent on many operational aspects but, clearly, time and temperature are the dominant factors. For example, modules operating intermittently at temperatures up to 1350°C have achieved a longer service life than similar grades used continually at only 1150°C.

Our experiences at these higher service temperatures have been mainly concerned with improving the performance of a large walking hearth, slab reheating furnace. The initial construction consisted of a bricked sidewall and roof. As in many other plants, early trials were made using 50mm thick ceramic fibre veneers cemented to the existing roof brickwork. Although some early savings were achieved, these were short lived as large areas of the fibre were lost (Figure 8). As part of the energy conservation programme, however, the decision was made to install a composite modular roof and replace the brick sidewall with a more thermally efficient monolithic structure, which was later veneered. The roof module was of novel design and represented a significant development at the time. The pre-assembled modules were 600mm x 600mm and had a total thickness of 280mm. The construction consisted of mineral wool, ceramic fibre blanket, a 26 grade insulating brick and at the hot face, 100mm of poly-crystalline alumina fibre (Figure 9). Performance of this module has been impressive and savings achieved considerable (Figure 10). Taking into account the developments described in the early part of this paper, specific energy consumption on this furnace has been reduced by more than 45%.

Trials are in hand to examine the potential of alternative fibres for use as the working lining. Some success has been achieved with the sidewall where various materials have shown promise, including ceramic fibre applied in the form of a spray. The roof, however, is a more difficult area and so far success has been limited.

The thermal performance of these materials can vary and some work is in hand to examine this aspect. Figure 11 illustrates the different temperatures recorded behind the 100mm thick working lining during one trial in which two different grades of fibre were being assessed. Although the temperature difference is not great, the potential savings are significant over the 100m² of the furnace roof.

5.   BENEFITS ARISING FROM THE USE OF CERAMIC FIBRES

Some of the benefits arising from the use of ceramic fibre have already been discussed. For the range of furnaces described, most of these benefits apply, but it is of interest to consider specific benefits which can be attributed to individual furnaces as shown in Table 3. It is important to stress that in order to achieve these benefits, production management must take full advantage of the gains made. This requires good communications and a committment to continually seek improvements in all aspects of production.

Clearly, it is difficult to accurately determine the contribution made to plant profitability by achieving improvements such as these, but it is worth stating that in British Steel, a 1% saving in energy yields a cost saving of approximately £9M/annum.

There is, clearly, a need for accurate data in order to compare performances, for the thermal benefits claimed by suppliers and installers are often not realised in service. We believe that this is an area which warrants further investigation wherever these materials are used. More information is needed in order to assess the actual performance of these materials in service for, as discussed earlier, the thermal performance of ceramic fibres can deteriorate dramatically in service, particularly when operating in difficult environments at the higher temperatures. The economics, therefore, need careful consideration.

When used successfully, ceramic fibres can achieve remarkable results with payback periods measured in months. In this respect they can often give a better return on capital expended than most of the other developments discussed. The main factors to consider with regard to the return on capital and how this affects the choice of refractory are, clearly, the capital cost of the installation and also the service life. The expected life of a fibre installation at British Steel Stainless is 5 years and any reduction on this is considered to be a failure. A minimum warranty of 12 months is required and we have successfully negotiated 24 months on several installations. In addition, the thermal efficiency of the installation is subject to a negotiated defects liability with the contractor.

In conclusion, however, although the current range of ceramic fibres includes materials which can give a satisfactory performance in many areas of application, a true high duty product, capable of operating at high temperatures under the more adverse conditions which prevail in some of our plants is still required.

When compared to the durability and performance of modern monolithic refractories, where largely maintenance free lives well in excess of 10 years can be confidently expected, ceramic fibres are clearly inferior but, as stated, can sometimes be more cost effective.

The thermal advantages of these materials, therefore, need to be carefully considered and as shown in this paper, these can be considerable. New ceramic fibre products are being developed and this work needs to be encouraged.

The main breakthrough will come, however, when materials are available which will tolerate the most difficult conditions and which will readily substitute for the more conventional lining materials, instead of, as at present, being limited to hot face insulation only.

ACKNOWLEDGEMENTS

The authors gratefully acknowledge the assistance given by colleagues within British Steel and also the managements of British Steel Stainless and British Steel Technical for permission to publish this paper.

REFERENCES

1. Hallet A G, Hay A C and Sheridan A T
    The Application of Self-Recuperative and Regenerative Burners in the Steel Industry
    J.Inst.E., 1987, 60 (3), 32-41

2. Hay A C
    Energy Saving in BSC Stainless
    Metals and Materials, January 1988

Lining	Relative Heat Stored at Equilibrium	Relative Heat Loss at Equilibrium
A. 300mm Firebrick	100	100
B. As A with 50mm Fibre Veneer	53	42
C. 175mm Mouldable 76mm Insulation	100	38
D. 255 Semi-Insulating Castable	57	38
E. As D with 50mm Fibre Veneer	41	25
F. 150mm Alumina Fibre 76mm Insulating Brick 63mm Fibre Blanket 45mm Mineral Wool	16	12
G. 200mm Fibre Module 50mm Fibre Blanket	6	15

STEADY STATE HEAT LOSS AND HEAT STORAGE CHARACTERISTICS OF ALTERNATIVE ROOF CONSTRUCTIONS

TABLE 1

Chemical Analysis, %	MINERAL WOOL	ALUMINO-SILICATE FIBRE				CRYSTALLINE FIBRE	
		STANDARD	HIGH DUTY	HIGH DUTY + Cr_2O_3	HIGH DUTY + ZrO_2	ALUMINA	MULLITE
Service Temperature °C	800-1000	1260	1400	1400	1400	1600	1600
Al_2O_3	16.0	46.2	52.1	43.1	34.9	95.0	72.0
SiO_2	46.0	53.2	47.5	54.3	49.3	5.0	27.0
Fe_2O_3	10.0	0.2	0.2	0.2	0.3	–	–
MgO	7.0	0.1	0.1	0.1	0.4	–	–
CaO	18.5	0.1	0.1	0.1	0.1	–	–
Cr_2O_3	–	–	–	2.2	–	–	–
ZrO_2	–	–	–	–	15.0	–	–

REFRACTORY FIBRE TYPES   TABLE 2

Type of Furnace	Approximate Operating Temperature °C	Major Benefits
Low Temperature Batch	<1000	Simple construction Cheap construction Light weight, lift off
High Temperature Batch	1100	Marked reduction in heating up times e.g. for one furnace reduced from 12h to 3h
Roller Hearth	1100	Hearth capacity increased by reducing wall thickness. Thermal performance maintained
Narrow Strip Annealing	1150	Increased output by elevating operating temperatures without increasing fuel costs
Wide Strip Annealing	1150	Improved strip temperature distribution following changes in furnace geometry
Slab Reheating	1350	Energy savings attributed to new ceramic fibre roof of 2 GJ/t

MAJOR BENEFITS ACHIEVED BY THE USE OF CERAMIC FIBRE LININGS

TABLE 3

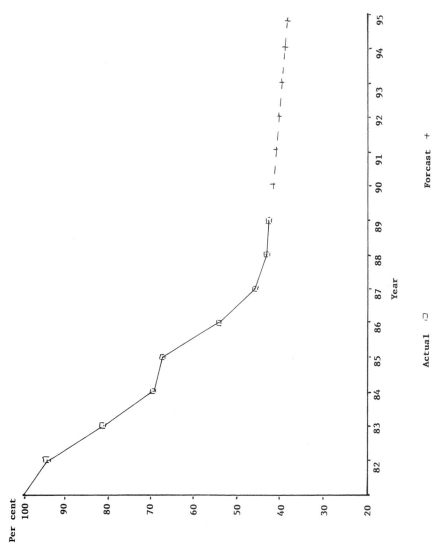

FLAT PRODUCTS ENERGY REDUCTION

FIGURE 1

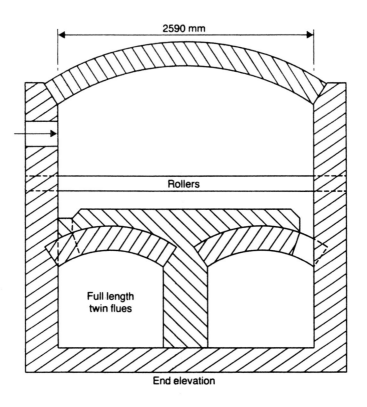

2590 mm

Rollers

Full length
twin flues

End elevation

ORIGINAL FURNACE LINING - ROLLER HEARTH FURNACE

FIGURE 2

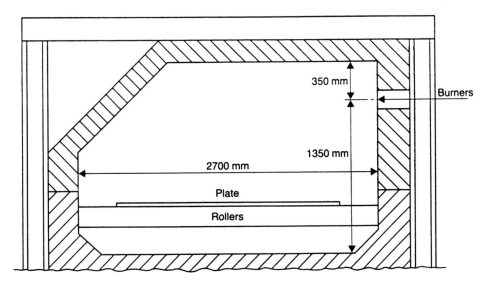

MODIFIED FURNACE LINING - ROLLER HEARTH FURNACE

FIGURE 3

LARGE HEAT TREATMENT FURNACE
CONTAINING 3.7m HIGH BELL

FIGURE 4

EXCESSIVE SHRINKAGE OF BLANKET LINING

FIGURE 5

INSTALLATION OF MODULAR LINING

FIGURE 6

OVERHEATED ANCHORAGE OF FAILED MODULE

FIGURE 7

VENEERED ROOF OF PLATE MILL FURNACE AFTER 20 WEEKS OPERATION

FIGURE 8

INSTALLATION OF COMPOSITE FIBRE MODULE ON PLATE MILL FURNACE ROOF

FIGURE 9

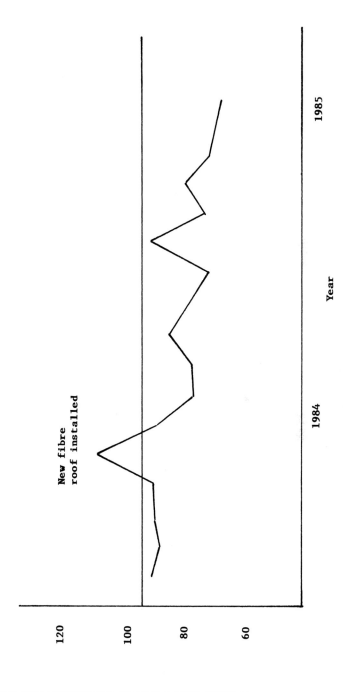

PERFORMANCE CHARACTERISTICS OF PLATE MILL FURNACE

FIGURE 10

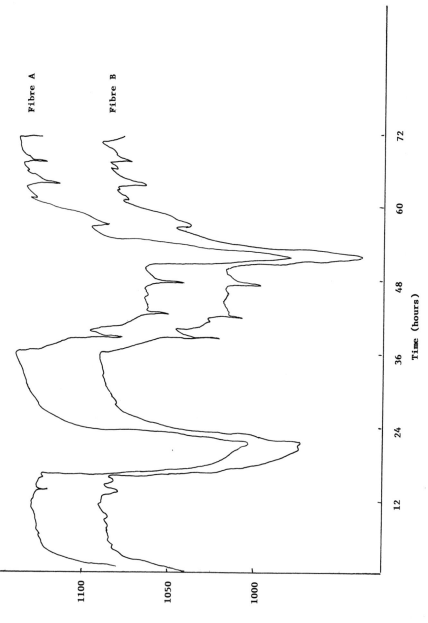

Fibre A

Fibre B

Temperature °C

1100

1050

1000

12   24   36   48   60   72

Time (hours)

REFRACTORY INTERFACE TEMPERATURE BEHIND 100mm THICK FIBRE VENEER

FIGURE 11

Utilization of new ceramics in industrial burners and equipment

A Hasesaka

Osaka Gas Co., Ltd.

ABSTRACT: As compared with heat resistant metals, some new ceramics have better forming property and dimensional accuracy than those of heat resistant alloy. This paper outlines the utilization of new ceramics for our industrial burners and equipments.

1. INTRODUCTION

In recent years, there has been a remarkable development of new ceramics available as structural material and heat resistance material.

In an attempt to utilize new ceramics for gas-fired industrial process heating and melting equipment, characteristics of new ceramics are compared with those of metal and are evaluated as follows:

Table 1: Evaluation of compared characteristics of new ceramics and metal

Advantage:	• Excellent heat resistance • Excellent oxidation resistance • Hard to react with molten metal
Disadvantage:	• Hard to be machined • Brittle to mechanical impact • Brittle to thermal shock • Lacks in reliability

As compared with heat resistant metals, some new ceramics have better forming property and dimensional accuracy than those of heat resistant alloy.

This paper outlines the utilization of new ceramics for our industrial burners and equipments.

2. EXAMPLES OF NEW CERAMICS USED IN INDUSTRIAL BURNERS

2-1 Ceramic Flame Holder

When a burner is subjected to preheated air at 600°C or over, the metallic flame holder may be damaged by heat.

A high alumina (Al_2O_3 = 76%) flame holder is mounted on a straight flame type burner developed for a crucible type aluminum melting and holding bath which uses preheated air of high temperature. The flame holder of a flat flame type burner developed for pot type glass melting furnace is also made from high alumina material.

A premix type flame holder has been developed to make a short flame which is fitted for a small size once-through boiler. The flame holder consists of many small alumina balls and the combustion reaction takes place in the narrow region above the surface of the holder.

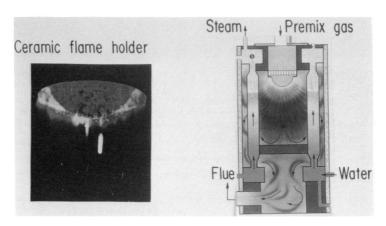

Fig. 1 Once-through Boiler Equipped with the Premix Ceramic Burner

2-2 Utilization as Burner Tiles

The burner of a heating furnace requires a combustion chamber in the burner. Because the combustion chamber is constructed with refractories, the chamber wall is thick and heavy. When a silicon nitride bonded SiC tube is used as the combustion chamber, a light and small size burner can be realized. Such burner is used for crucible type copper melting furnace whose furnace temperature is 1,400°C or more.

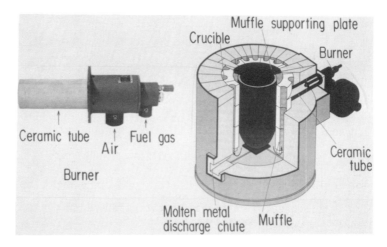

Fig. 2 Burner Having Ceramic Tube Combustion Chamber and
        Copper Melting Furnace Equipped with the Burner

2-3 Burner of Ceramic Fiber

A silicate alumina fiber matrix is being developed jointly with a manufacturer for use as the combustion plate for a large infrared rays burner (surface temperature of 850°C).

In case of the catalytic combustion burner whose surface temperature is 400°C, a catalyst is held in the porous alumina long fiber mat.

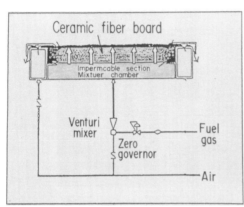

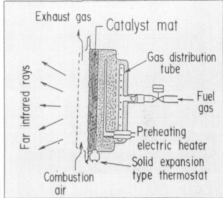

Fig. 3   Large Scale Infrared Rays   Fig. 4   Catalytic Combustion
          Burner   Burner

3.   EXAMPLES OF NEW CERAMICS USED IN INDUSTRIAL EQUIPMENTS

3-1   Ceramic Tube Type Forging Furnace

A ceramic tube type forging furnace is put to use, in which a round bar is inserted into a silicon nitride bonded SiC tube and the outside of the tube is heated with a gas burner. This makes possible of heating with less scale loss.

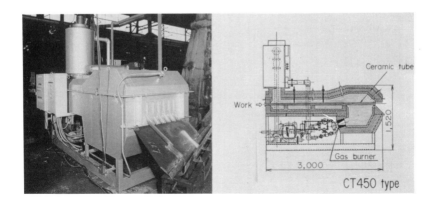

Fig. 5 Ceramic Tube Type Forging Furnace

3-2 Immersion Tube Type Aluminum Holding Furnace

A single ended type ceramic immersion tube heater is developed for aluminum holding furnace for aluminum die casting. Tube materials are sintered Si_3N_4 or silicon nitride bonded SiC.

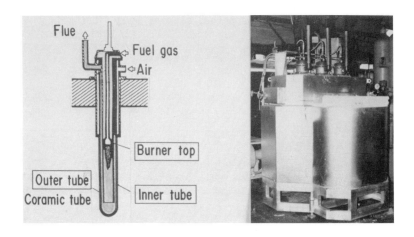

Fig. 6 Ceramic Tube Immersion Heater for Aluminum Holding Bath

3-3 Ceramic Heat Exchanger

A heat exchanger is developed which employs a small matrix type heat exchanging element made from cordierite having good forming property which is not available by heat resistant metal.

When mounted on a crucible type aluminum melting and holding bath, highly efficient heating at preheated air temperature of 600°C is achieved. A multi-tube type heat exchanger is under way for development.

Fig. 7 Matrix Type Ceramic Heat Exchanger

3-4 Far Infrared Rays Heater

A far infrared rays heater is put to use, in which ceramics such as ZrO_2 which has small emissivity on the short wavelength side and the emissivity on the long wavelength side is close to black body is coated on a metallic tube. In the metallic tube gas is combusted to achieve the surface temperature of about 400°C.

This far infrared rays heater is used for paint baking dryer and resin foaming furnace.

4. CONCLUSION

In addition to applications of ceramics stated above, ceramic radiant tube and ceramic recuperative burner are under research and development.

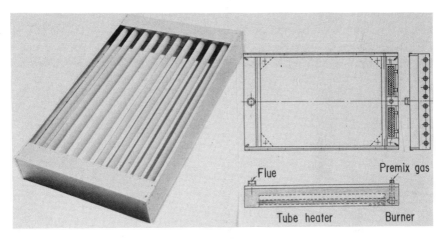

Fig. 8 Far Infrared Rays Heater

We, Osaka Gas, continue to utilize new ceramics actively while we develop industrial gas equipment so that energy saving and high quality of processed products can be accomplished.

Silicon carbide in energy applications

Peter Roebuck and Wolfgang Heider*

Kanthal Limited, Inveralmond Industrial Estate, Perth, PH1 3EE
(*Kanthal GmbH, Postfach 147, Aschaffenburger Strasse, D-6082 Mörfelden-Walldorf, West Germany)

ABSTRACT : With the ever growing demand for high performance materials, particularly in the ceramic industry for components such as kiln cars and rollers,a new range of silicon carbide (SiC) materials has been developed.The outstanding properties of these materials make them ideal for high temperature applications. SiC helps to reduce the refractory mass in traditional furnaces and is essential in Fast-Fired furnaces where it's high strength and thermal shock resistance improve the wares-to-kiln furniture ratio,which in turn reduces the energy consumption. The thermal and mechanical properties and the design methodology are discussed in detail.

1. INTRODUCTION

The ceramic industry is one of the most important industries of the industrialised countries. The manufacture of table and sanitary ware as well as of technical ceramics belongs to one of the branches with, as shown by Hoffmann *et al* (1980) the highest energy consumption, especially for drying and firing these goods at high temperatures. From the energy consumption point of view, refractories play an important role as they are essential for transporting and supporting the ceramic ware through the furnace. When tunnel kiln furnaces were first introduced, the relationship between the weight of the refractories to that of the ware to be fired was ten to one i.e. most of the energy was necessary to heat the refractories rather than the ware. During the last decade, the use of more economical fast firing processes has increased, processes for which special refractories with high hot strength, excellent thermal shock resistance, low mass and low energy consumption are essential. It is for these reasons that refractories made of different grades of silicon carbide have been developed which are capable of meeting the technical demands of modern fast firing processes. The use of these materials make the firing stage both more efficient and economical.

In the following sections the possibilities for energy savings using the recently introduced RSiC and SiSiC refractories with outstanding thermal mechanical properties as well as newly developed manufacturing techniques are described.

2. MANUFACTURING

The first refractories were made of specially selected fire clays and had poor thermal and mechanical properties as well as a short service life.The second generation of refractories were made of clay-bonded silicon carbide and cordierite which had better thermal shock resistance and higher strength.

Silicon carbide is by no means a new material, having been widely used for the past 80 years as a high duty refractory, as an abrasive and as a heating element. This company has itself been associated with the manufacture of high quality SiC heating elements for over 30 years.

Self-bonded silicon carbide grades such as recrystallised RSiC and silicon infiltrated SiSiC are the latest generation of advanced ceramics and are able to optimise all kinds of firing processes. The manufacturing route and the design are important parameters when considering using these latest refractories in particular applications. The first commercial RSiC refractories were produced by slip-casting which is ideal for small and complex shaped articles. For larger articles and those of simpler design such as beams and rollers, the use of extrusion as a forming route has been shown to be preferable (Willmann and Heider 1981) to all other shaping methods. That is why both techniques were developed for manufacturing our new SiC refractories as shown schematically in Figure 1.

These new SiSiC and RSiC materials are known as ALPHALITE 1300 and ALPHALITE 1600 respectively.

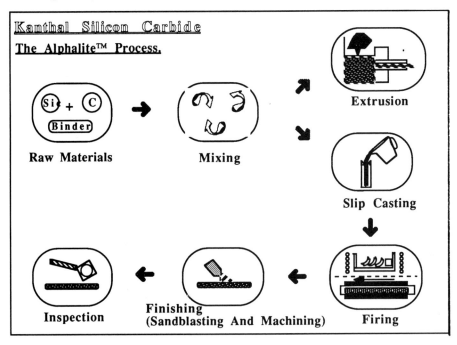

Figure 1   Schematic Representation of the
              Manufacture of RSiC and SiSiC Refractories

The new manufacturing route helps to extend the shaping capability, improve product quality and to reduce the manufacturing costs.

3. PROPERTIES

For a refractory material to be used as a kiln car superstructure, it must satisfy the following product requirements:

* high operating temperature
* high loading capacity
* no sagging under load
* no chemical reaction with ware being fired
* no chemical reaction with firing atmosphere
* high degree of flexibility
* easy to handle and replace
* low mass
* low energy consumption
* long service life
* low manufacturing

Self-bonded silicon carbide meets these requirements.

Table 1 shows the mechanical,thermal and chemical properties of RSiC and SiSiC in comparison with other refractory materials, from which their suitability can be clearly seen.

Material	PROPERTIES						
	T.Max	Density	Strength	CTE	Specific Heat	Thermal Cond.	Thermal Shock
	deg C	g/cc	M Pa	10^{-6}/K	J/kgK	W/mK	
SiSic	1350	3.0 d	300	4.4	1050	50	**
RSiC	1650	2.8 p	100	4.6	1200	30	*
Oxide - SiC	1450	2.5 p	20	5.0	1090	8	0
Cordierite	1280	2.0 p	15	2.9	1500	1	**
Chamotte	1450	2.1 p	4	5.3	1000	1	-
Mullite	1700	2.7 p	10	6.7	1040	1.5	-
Corundum	1750	2.7 p	6	6.2	1170	2	--

d = dense   p = porous
Table 1   Properties of Various Refractory Materials

The most important factors for the use of both SiC refractories are the mechanical strength (at both room and elevated temperature), thermal properties and oxidation resistance.Each of these will now be discussed in more detail.

3.1 Strength as a Function of Temperature

The strength of ceramic materials is usually measured by using the three or four point bending method at room temperature on small test bars.As the refractories are normally used at temperatures above 1000 deg.C,it is important to know the strength/temperature relationship in order to optimise the design of a high loading carrying system.Figure 2 shows the bending strength as a function of temperature for different refractory materials.

The silicon infiltrated silicon carbide exhibits the highest strength whilst that of the recrystallised silicon carbide,though lower,remains constant throughout the whole range of application temperatures due to the absence of any glassy phase.All the other refractory materials such as cordierite and the alumino-silicates have a glassy phase which starts to soften as the temperature increases.

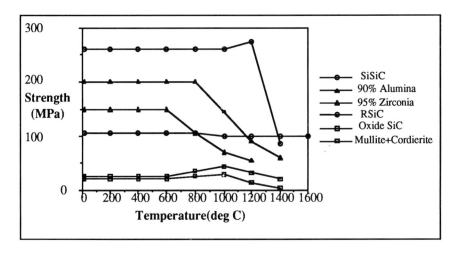

Figure 2   Strength as a Function of Temperature for Various Materials

The drastic and rapid reduction in hot strength of SiC is due to the melting of the excess pure silicon.in it's structure.Up to this temperature there is a slight increase in strength.caused by surface oxidation and by removal of surface defects.

3.2 Strength as a Function of Time

It is well known that the structure of ceramics,which has a strong influence on the materials strength,changes as a function of time and temperature.In the case of alumino-silicates such as sillimanite,the strength can be greatly reduced under certain conditions of time and temperature. That is why such refractories can fail under load at high temperatures,thereby risking damage to the ware and incurring expensive furnace repairs.

The strength/time relationship for our silicon carbide grades has therefore been investigated in detail (see Figure 3).The recrystallised grade of silicon carbide,with an open porosity of 12 to 14%,showed little reduction in strength when tested at 1400 deg.C after being heat treated at 1000 deg.C for a period of 1000 hours.Under similar conditions, other commercially available RSiC grades showed much greater and significant reductions in strength caused by oxidation and the formation of a silica (SiO_2) layer along the grain boundaries.The reason for the superior performance of the new RSiC type is the reduced pore volume and the smaller pore sizes which prevent the internal oxidation of the structure, especially at the lower temperature range between 800 and 1200 deg.C.The oxidation behaviour of the SiSiC fully dense type is characterised by the formation of a dense silica layer on the outer surface which has no effect on the materials strength throughout it's normal range of operating temperatures.

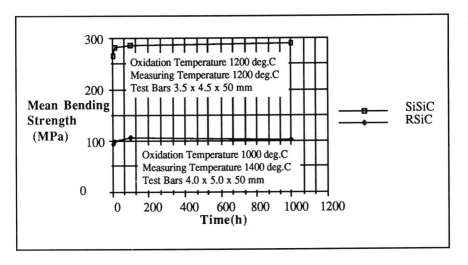

Figure 3   Strength of SiSiC and RSiC Alphalite
            Materials as a function of Oxidation Time.

3.3 Strength as a Function of Volume

The volume effect on strength depends upon the brittleness and the flaw size distribution, as described by the m-parameter of the Weibull statistics based on Griffith's weakest-link hypothesis .This theory was originally proposed by Weibull (1951) and subsequently discussed by Bansal (1976), Lewis and Oyler (1976), Kleer *et al* (1986), Rice (1987) and Buresch and Meyer (1988).

Whilst the volume effect of all the traditional oxide refractories remain for the most part unknown, this effect has been studied in detail for both RSiC and SiSiC.Figure 4 shows the calculated strength values based on the formula:

$$\sigma_2 = \sigma_1 \ (V1/V2)^{-1}/m$$

(where σ_2 = 4 point bending strength of specimen 2 in N/mm^2

σ_1 = 4 point bending strength of specimen 1 in N/mm^2

V1 = volume of specimen 1 in mm^3

V2 = volume of specimen 2 in mm^3

m = Weibull modulus)

Compared with experimental data on SiSiC tubes the difference between the experimental and predicted strength values has been attributed to the sample surface conditions.When the data from the small specimens are extrapolated for larger components, unrealistic values are produced.Similar investigations are being carried out on RSiC as well and those results will be reported in due course.

Once these investigations are completed,we should be able to produce an actual SPT diagram, up until now available for SiSiC only from the work of Richter *et al* (1982), for high temperatures and varying volumes which will replace the presently used safety factors.

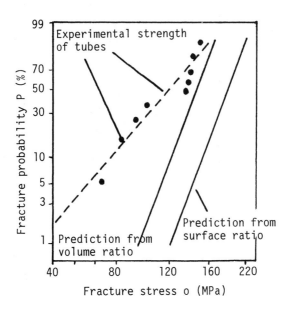

Figure 4   Strength as a Function of Volume for SiSiC

3.4 Oxidation Behaviour

Because refractories are normally used in oxidising atmospheres,their resistance to oxidation is a very important property. In the case of porous materials like RSiC, the reaction with the air has a large effect on the thermal mechanical properties and on the service life. In fact the oxidation rate depends on several different parameters such as temperature,time, pore volume and size distribution as well as impurities in the material and the manufacturing process used.

Figure 5 shows the weight gain of some different silicon carbide grades such as RSiC and Si_3N_4-bonded SiC as a function of time and temperature. From the graph it can be seen that the weight increase is actually lower at the higher temperatures.This is because the initial oxidation at low temperatures forms silica inside the pore structure.At temperatures above 1200 deg.C a dense silica layer forms on the outer geometrical surface allowing further oxidation only at a slow rate by diffusion of oxygen through this silica layer.If the pores are completely filled with silicon metal as in the case of SiSiC,the oxidation rate is very low, limited at higher temperatures again by the rate of oxygen diffusion through the silica layer as shown by Schlichting (1979) and Förthmann and Naoumidis (1989).

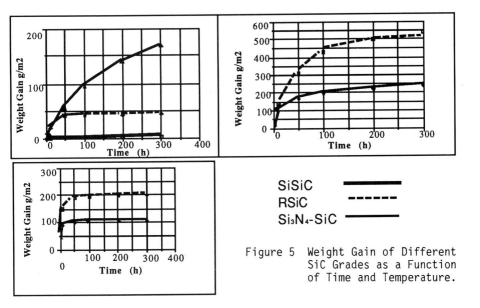

Figure 5   Weight Gain of Different SiC Grades as a Function of Time and Temperature.

3.5 Thermal Properties

It is an established fact that most ceramic materials are good thermal insulators possessing poor thermal shock resistance in combination with a high degree of brittleness and low strength. Figure 6 shows the thermal conductivity of RSiC and SiSiC as measured by the laser flash method,in comparison with other refractory materials at 1000 deg.C. The thermal conductivity of SiSiC is 50 W/mK,a value between 20 and 50 times higher than conventional oxide ceramics.The porous recrystallised silicon carbide also has a high value for thermal conductivity though lower than that for SiSiC. In addition to these high values, both SiC grades possess a low coefficient of thermal expansion and as already discussed a high hot strength; a combination of properties giving rise to outstanding thermal shock resistance.This latter parameter is very important for all fast firing processes where the cycling times from room temperature to 1400 deg.C are of the order of only 4 hours.

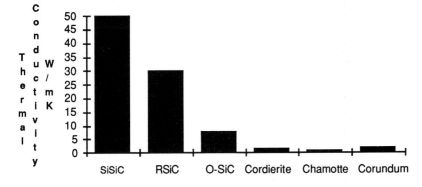

Figure 6   Thermal Conductivity of RSiC and SiSiC

Initial measurements of the thermal fatigue of SiSiC show no strength degradation up to 600 cycles between 100 and 1200 deg.C at a cooling rate of 80 deg.C per second as shown by Röttenbacher and Heider (1988). Similar measurements are being carried out on RSiC,the results of which will be reported in due course.

4. DESIGN METHODOLOGY

The combination of the outstanding properties of both the silicon carbide grades already discussed together with the forming capabilities of the newly developed manufacturing techniques makes possible the production of refractories of improved design capable of meeting the necessary technical and economical requirements. These requirements could not be met before by the use of traditional refractories.

The following example refers to the design of a low mass carrying component capable of improving the firing process by increasing the furnace capacity and reducing the energy consumption. Using the new manufacturing method, rectangular beams (as traditionally used in tunnel kilns as well as fast firing kiln furnaces) can be extruded with thin walls (2.7 to 8.0mm) in large cross sections and long lengths. The main advantage of this technique is that not only are the walls thin but also very uniform with tolerances of less than ± 0.2mm. Such beams have low mass and excellent thermal shock resistance.

The load carrying capacity of SiSiC beams of different cross sections as a function of span width (1000 to 2500mm) is shown graphically in Figure 7. The calculations are based on a uniform stress over the whole length and applying a safety factor of 5 to include the statistical nature of the materials properties as well as the temperature, time and volume effect on strength.

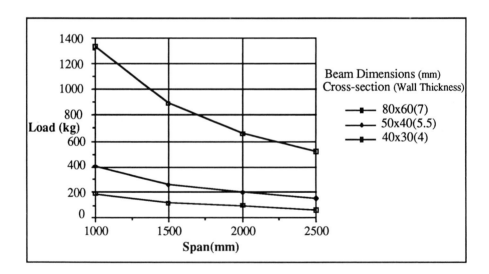

Figure 7   Load Carrying Capacity of SiSiC Beams

In Figure 8 the load carrying capacity of SiSiC beams are compared with those of RSiC and cordierite for a constant cross section of 80mm x 60 mm. For these 3 beam materials, the load carrying capacity can be seen to be 1330kg, 550kg and 80kg respectively. For the cordierite beams the calculations are based on a density value of 1.95g/cc, a mean bending strength of 15MPa and a safety factor of 5.

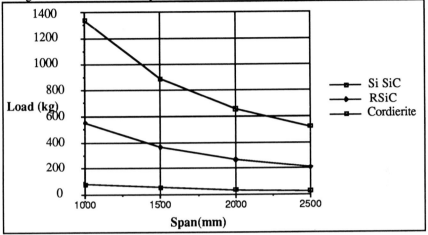

Figure 8   Comparison of Load Carrying Capacities of SiSiC,RSiC and Cordierite

For a constant load of 1330kg across a 1000mm span, different cross sections and wall thicknesses are required for the various refractory materials, and these are shown in Table 2. These different cross sections give weights for the beams of 25.5kg (cordierite), 9.3kg (RSiC) and 5.1kg (SiSiC), as shown in Figure 9.

Span	Spec. Load	Size of Beam (Cross Section x Wall Thickness) (mm) and weight		
mm	kg	SiSiC	RSiC	Cordierite
1000	1333	80 x 60 x 7 5.1	100 x 80 x 10.6 9.3	175 x 130 x 30 28.6
1000	406	50 x 40 x 5.5 2.6	70 x 50 x 7 4.1	120 x 80 x 24 14.2
1000	179	40 x 30 x 4 1.4	50 x 40 x 6.2 2.7	90 x 60 x 22 9.1

Table 2   Comparison of Sizes of Beams Required to Carry a Specific Load

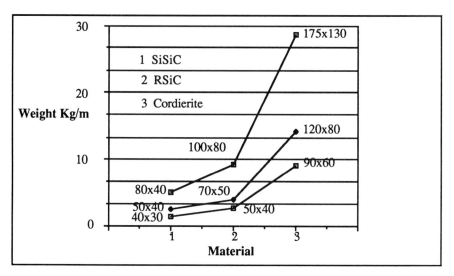

Figure 9   Comparison of Weights of Beams of Various Materials at Constant Load

A similar relationship exists for energy consumption (see Figure 10) calculated by the formula:

$Q = m \times c \times \Delta T$
(where Q= Energy consumption in KJ, m= mass in kg, c= specific heat in kJ/kgK and ΔT = Temperature difference)

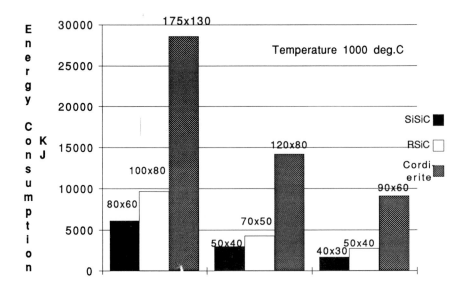

Figure 10   Comparison of Energy Consumption of SiSiC, RSiC and Cordierite Beams

For cordierite beams the energy consumption is 5 times higher than that for SiSiC, and that with a bulk density only 40% that of SiSiC. The difference in energy consumption between RSiC and SiSiC is only small since the larger RSiC cross section is partly compensated for by it's lower bulk density and specific heat.

It can therefore be seen that it is possible to significantly reduce the energy consumption by using self-bonded silicon carbide instead of traditional oxide ceramics, particularly in modern fast firing furnaces.

5 APPLICATIONS

Two case histories will be discussed to illustrate the advantages of SiSiC and high density RSiC materials as produced using the previously mentioned manufacturing techniques.

The first example describes the use of SiSiC (Alphalite 1300) rollers in a modern fast firing roller hearth kiln furnace. The rollers were 40mm diameter x 2500mm long with a maximum deflection of less than 2mm over the full length in an air atmosphere and furnace temperature of 1180 deg.C. Maximum load was 15 kg. The rollers were removed over a period of weeks to check for any bowing and to see the extent of surface depositions. The first roller was removed after 35 days in operation, a further 2 after 54 days and 2 more after 89 days. They all performed very well, maintaining their straightness, surviving the thermal shock of quick removal from the furnace (unlike the normally used sillimanite rollers which have to be withdrawn from the kiln very carefully and very slowly because of their poor thermal shock resistance) and with a minimal amount of surface deposition. This surface layer was very flat, did not disturb the conveying of ware and needed to be removed by cleaning with conventional grit at periods of about 4 to 6 months. The cleaning of the sillimanite rollers has to be done at 3 monthly intervals,and has to be done by centreless grinding. This removes material from the outer wall and hence can only be done a limited number of times before the reduced wall thickness and strength will lead to breakage.

Other points to note are that the safe working load of the Alphalite 1300 beam is 5 to 8 times higher than that of sillimanite, and that an Alphalite roller can be changed in about 10 minutes with a much reduced production downtime compared with the 1 hour required to change the sillimanite roller, a longer service life (due to retained hot strength) and an increase in the kiln capacity.

The second example is of the use of RSiC(Alphalite 1600) beams for light weight kiln furniture car construction in the firing of table ware. The test conditions were as follows:

```
Type of furnace: tunnel kiln furnace 110m long
Carrying system: single deck kiln furniture cars 2200 x 1100mm
Furnace temperature:      1400 deg.C ± 20 deg.C
Atmosphere:               up to 1180 deg.C -oxidising
                          1100 to 1400 deg.C -reducing
                          1400 to room temperature-oxidising
Cycling time: 26 hours
Type of beams: currently use mullite 50 x 100 x 1080mm
Tested material:   Alphalite 1600   40 x 40 x 1080 (wall thickness 4.5mm) and
                   alternative traditional RSiC slip cast 50 x 40 x 1080mm(variable
                   wall thickness of 7 to 8mm)
 Number of beams: 8 per kiln car
 Mechanical load:   1200kg(total)
```

The expected lifetime for the RSiC beams is 250 cycles minimum (3 to 4 cycles per week). The trial has only been in progress since September 1989. The long term performance will be reported in due course. Table 3 shows the comparison of weight and energy consumption between mullite, RSiC slip cast and RSiC extruded (Alphalite 1600) materials.

Material	Sizes (mm)	Weight (kg)	Number of pcs.	Total Weight kg	Energy Consumption kJ per beam
Mullite	100 x 50 x 1080	11.5	8	92.0	128,800
RSiC Slip Cast	40 x 50 x 1080	3.2	8	25.6	37,630
Alphalite 1600 (extruded RSiC)	40 x 40 x 1080	1.9	8	15.1	22,200

Table 3   Comparison of Beams for Firing of Table Ware

The advantages of Alphalite beams in comparison with mullite and slip cast RSiC beams can be summarized as: lower mass, higher strength, better oxidation resistance due to the higher density, excellent thermal shock resistance due to the uniform wall thickness, and a predicted longer service life.

6 REFERENCES

Bansal G.K. *et al:* 1976 J. Amer. Ceram. Soc. 59 No. 11-12, p.472.
Buresch F.E., Meyer W: 1988 Materialprüfung 30, No.6, p.205.
Förthmann R., Naoumidis A., 1989 High temperature corrosion of technical ceramics
   Petten, Netherlands.
Hoffmann U. *et .al:* 1980 Ber. Dt. Keram. Ges. 57 No.2.
Kleer G. *et.al:* 1986 Ceram. Mat. and Comp. for engines. Lübeck, Germany p.979.
Lewis III, D., Oyler S.M.: 1976 J.Amer. Ceram.Soc. 59 No.11-12, p.507.
Rice R.W.: 1987 J.Amer. Ceram. Soc. Bull. 66 No.5, p.794
Richter H. Willmann G. Heider W.: 1982 Z.F. Werkstoff-Technik 13, p.355.
Röttenbacher R., Heider W., 1988 Im Technische Keramik, Vulkan Verlag, Essen,
   p.214.
Schlichting J: 1979 Ber.Dt. Keram. Ges. 56 p. 196 and 256.
Weibull W.: 1951 J. Appl. Mech. 18, p.293
Willman G., Heider W. 1981 Sprechsaal 13.

Author Index

Subject Index